Monohydric Alcohols

Monohydric Alcohols

Manufacture, Applications, and Chemistry

Edward J. Wickson, EDITOR
Exxon Chemical Company

Based on a symposium

sponsored by the Division of

Industrial and Engineering Chemistry

at the 179th Meeting of the

American Chemical Society,

Houston, Texas,

March 25–26, 1980.

ACS SYMPOSIUM SERIES 159

AMERICAN CHEMICAL SOCIETY
WASHINGTON, D. C.   1981

Library of Congress CIP Data

Monohydric alcohols.
  (ACS symposium series 159; ISSN 0097-6156)

  Includes bibliographies and index.

  1. Alcohols—Congresses.
  I. Wickson, Edward J., 1920-   II. American
Chemical Society. Division of Industrial and Engi-
neering Chemistry. III. Series.

TP248.A5M66   661'.82   81-5950
ISBN 0-8412-0637-6   AACR2
                  ASCMC 8 159 1–222 1981

ACS Symposium Series

M. Joan Comstock, *Series Editor*

FOREWORD

The ACS Symposium Series was founded in 1974 to provide
a medium for publishing symposia quickly in book form. The
format of the Series parallels that of the continuing Advances
in Chemistry Series except that in order to save time the
papers are not typeset but are reproduced as they are sub-
mitted by the authors in camera-ready form. Papers are reviewed under the supervision of the Editors with the assistance
of the Series Advisory Board and are selected to maintain the
integrity of the symposia; however, verbatim reproductions of
previously published papers are not accepted. Both reviews
and reports of research are acceptable since symposia may
embrace both types of presentation.

CONTENTS

Preface ix

1. **Chemistry of Monohydric Alcohols** 1
   F. M. Benitez

2. **Methanol: Manufacture and Uses** 19
   T. F. Kennedy and D. Shanks

3. **Methanol from Wood in Brazil** 29
   V. Brecheret, Jr. and A. J. A. Zagatto

4. **Ethanol: Manufacture and Applications** 47
   I. B. Margiloff, A. J. Reid, and T. J. O'Sullivan

5. **Ethanol in Motor Gasoline** 55
   T. Tarr and J. R. Jones

6. **Manufacture of *n*-Butanol and 2-Ethylhexanol by the Rhodium Oxo Process and Applications of the Alcohols** 71
   C. E. O'Rourke, P. R. Kavasmaneck, and R. E. Uhl

7. **Manufacture of Higher Straight-Chain Alcohols by the Ethylene Chain Growth Process** 87
   P. H. Washecheck

8. **Applications of Higher Alcohols in Household Surfactants** 101
   T. P. Matson

9. **Secondary Alcohol Ethoxylates: Physical Properties and Applications** 113
   N. Kurata, K. Koshida, H. Yokoyama, and T. Goto

10. **Higher Linear Oxo Alcohols Manufacture** 159
    R. E. Vincent

11. **Synthetic Lubricant Basestocks from Monohydric Alcohols** 165
    J. B. Boylan, B. J. Beimesch, and N. E. Schnur

12. **Monohydric Alcohol Ester Plasticizers for Polyvinyl Chloride: Past, Present, and Future** 181
    E. J. Wickson

13. **Monohydric Alcohols in the Flavor and Fragrance Industry** 197
    J. Dorsky

Index 213

PREFACE

Unsubstituted monohydric alcohols are the largest single class of oxygenated hydrocarbons made in the chemical industry. In fact, three of the alcohols—methanol, ethanol, and isopropyl alcohol—are in *Chemical and Engineering News* "Top 50 Chemicals" for 1979. These alcohols are used in a myriad of applications, some very large and others quite small. Solvents, fuels, antifreeze, germicides, frothers, antifoams, and inks are examples of where the alcohols are used as such. Derivatives cover an even wider range of applications, including resins, solvents, plasticizers, surfactants, lubricants, flavors, fragrances, capacitor fluids, and the fast growing gasoline additive MTBE, to name a few. In this sense, alcohols differ markedly from some of the very large volume organics such as ethylene, propylene, toluene, *p*-xylene, and urea, which are used in very large volumes but have relatively few end uses.

It is common practice to divide alcohols into lower alcohols (C_1–C_5) and higher alcohols (C_6 and higher). In turn, higher alcohols are generally subdivided into (polyvinyl chloride) plasticizer range (C_6–C_{11} or C_6–C_{12}) and detergent range (C_{11} or C_{12} and higher). Actually, the dividing line is not this clear cut, and one finds the lower alcohol *n*-butanol used in butyl benzyl phthalate and the detergent-range tridecyl alcohol used in ditridecyl phthalate. Both of these esters are important polyvinyl chloride plasticizers. Also, detergents based on mixed alcohols, including C_{11}, are produced commercially, and there are many surfactants, such as di-2-ethylhexyl sulfosuccinate, based on the so-called plasticizer-range alcohols.

In developing the symposium on which this book is based, our objective was not to cover all monohydric alcohols, but to cover representative lower and higher alcohols with special attention, when possible, to new developments and applications.

The book starts with a review of the chemistry of monohydric alcohols with references to many of the common reactions used in industry.

Much attention has been devoted in the professional journals and in the press to gasohol, and announcements of new ethanol plants and process improvements seem unending. A perspective on the problems and promise of gasohol is covered along with a status report on work in Brazil on (fuel) methanol from wood, not by destructive distillation used in the past, but by modern synthetic chemistry. Other promising new uses for methanol are covered that together suggest enormous growth for this first member of the monohydric alcohol series.

A detailed discussion on surfactants from secondary alcohols which are relatively little known in the U.S. is included, together with a review of linear alcohol processes (Oxo and Ziegler) and detergent applications of the Ziegler alcohols. Also covered is a discussion of the revolutionary rhodium oxo process which has already resulted in a number of new plants—announced, under construction, or in operation, worldwide—for the manufacture of *n*-butanol and 2-ethylhexanol. Applications of these alcohols are also discussed.

The application of branched-chain alcohol diesters in part-synthetic lubricants is covered as one of the ways the petroleum industry is responding to the challenges brought on by the trend toward new smaller cars with more demanding lubricant requirements.

The history of monohydric alcohol–based ester-type plasticizers for polyvinyl chloride is reviewed, and a prediction of the future for these plasticizers in the eighties is made. Finally, although not in the large volume category compared to many monohydric alcohols, the rather sophisticated chemistry used to produce monohydric alcohols for the flavors and fragrance industry is covered.

EDWARD J. WICKSON
Exxon Chemical Company
P.O. Box 241
Baton Rouge, Louisiana 70821

April 3, 1981

Chemistry of Monohydric Alcohols

F. M. BENITEZ

Exxon Chemical Company, Specialties Technology Division, P.O. Box 241,
Baton Rouge, LA 70821

Alcohols can be regarded as hydroxyl derivatives of hydrocarbons. They can be characterized by the number of hydroxyl
groups (monohydric, dihydric, etc.), according to their structure (primary, secondary or tertiary), and by the structure of
the hydrocarbon function to which the hydroxyl is attached
(aliphatic, cyclic, saturated or unsaturated).

This chapter is concerned almost exclusively with the
chemistry of saturated aliphatic monohydric alcohols with
particular emphasis on the reactions used in the conversion of
these alcohols to other useful compounds. Manufacture of many
of the alcohols is covered in other chapters.

Acidity and Basicity

Alcohols are amphoteric and thus can function both as weak
Brønsted acids and as bases:

$$R\text{-}OH + Z^{\ominus} \rightleftharpoons R\text{-}O^{\ominus} + ZH \quad (ROH \text{ acting as an acid})$$

$$R\text{-}OH + HA \rightleftharpoons ROH_2^{\oplus} + A^{\ominus} \quad (ROH \text{ acting as a base})$$

The acidity of the hydroxyl group can be seen in the rapid
proton-deuteron exchange that can take place when alcohols are
dissolved in D_2O (Reaction I), alkali metals (Reaction II) and
organometallic reagents (Reaction III and IV).

$$CH_3CH_2OH + D_2O \rightleftharpoons CH_3CH_2OD + DOH \tag{I}$$

$$CH_3OH + Na \longrightarrow CH_3O^{\ominus}Na^{\oplus} + 1/2\ H_2 \tag{II}$$

$$CH_3CH_2CH_2OH + CH_3Li \longrightarrow CH_3CH_2CH_2O^{\ominus}Li^{\oplus} + CH_4 \tag{III}$$

$$CH_3CH_2CH_2OH + C_3H_7MgBr \longrightarrow CH_3CH_2CH_2O^{\ominus}MgBr^{\oplus} + C_3H_8 \tag{IV}$$

The most complete measurements of the acidity of alcohols in water were made some time ago by Long and Ballinger (1,2,3) using conductivity methods. The pKa values for substituted methanols (RCH_2OH) (2) are a linear function of the Taft $\sigma*$ constants (4,5) for the R substituents, allowing the prediction of the actual pKa by using the formula: pKa = 15.9 - 1.42 $\sigma*$.

In recent years the question of acidity and basicity has been reopened by the development of techniques to measure them in the gas phase(6). The results available reemphasize the fact that solvation factors have a profound influence on the course of acid-base reactions.

Brauman and Blair have determined (6) that the acidity of some substituted alcohols increases as the size and number of substituents increase. This is exactly the opposite effect seen in solution measurements. The conclusion that must be deduced from this is that there are two kinds of acidity that must not be confused: a) an intrinsic acidity, which is best approximated by gas-phase measurements and which reflects the properties of the ions and the molecules in isolation, and (b) a practical liquid phase acidity in which solvation effects play a very important role. In the interpretation of structure-acidity relationships in solvents, the results will probably be misleading unless the structures being compared are very similar.

Categories of Chemical Reactions of Alcohols

The following sections on the chemical reactions of alcohols has been broken down into five categories: (A) Nucleophilic reactions of alcohols, (B) displacement of the hydroxyl group, (C) dehydration of alcohols, (D) oxidation of alcohols, and (E) analytical determination of the hydroxyl group.

Under each one of these categories the different types of reactions are organized in a logical manner. Some examples are given, but by no means are all the different types of alcohols covered. The reader is asked to extend the analogies and use the references given to pursue his areas of interest.

Nucleophilic Reactions

Any species having an unshared pair of electrons may act as a nucleophile, whether it is neutral like an alcohol or negative like the alkoxide ion. The rate of S_N1 reactions is independent of the structure and charge of the nucleophile. For S_N2 reactions, factors like the charge of the nucleophile, its degree of solvation and nucleophilicity determine the rate of the reaction (6A).

The major trend in nucleophilicity is to parallel base strength. However, sometimes differences between basicity and nucleophilicity of a species occur because the two are somewhat different. Basicity measures attack on hydrogen and it is

thermodynamically controlled. Nucleophilicity on the other hand
measures attack on carbon and it is kinetically controlled.

   $\underline{\text{Reactions of the Alkoxide Ion}}$. The nucleophilic reactions
of the alkoxide ion (RO^-) are very similar to those of the
hydroxide ion (OH^-) with the exception that the latter has an
extra ionizable proton which can lead to further reaction after
the initial nucleophilic attack.
   The displacement of the bromide ion from an alkyl bromide
(Reaction V) by an alkoxide has been found to be a first order
reaction in both the reactant and substrate ($\underline{7}$). It is implicit

$$RO^{\ominus} + R'-Br \longrightarrow ROR' + Br^{\ominus} \tag{V}$$

in the reaction above that inversion of configuration will take
place at the alkyl halide. The nucleophilic displacement of a
halide by an alkoxide is commonly known as the Williamson ether
synthesis and is still the best general method for the preparation of symmetrical and unsymmetrical ethers (7A).
   The transformation of chlorohydrins into the corresponding
epoxides (Reaction VI) may be regarded as a special case of the
Williamson reaction. Many epoxides have been made this way

$$-\overset{|}{\underset{|}{C}} - \overset{|}{\underset{|}{C}} - \quad \xrightarrow{\quad OH^{\ominus} \quad} \quad - \overset{|}{C} --- \overset{|}{C} - \; + \; NaCl \tag{VI}$$
$$\;\;\; Cl \;\; OH \qquad\qquad\qquad\qquad \diagdown_O\diagup$$

and the method is generally useful for the synthesis of five and
six-membered rings. There is a large amount of evidence for an
intramolecular mechanism ($\underline{8}$).
   When the halide is bonded to an allylic system (CH_2=CH–CH_2–X)
an alkoxide ion will react analogously to the previously described
S_N2 displacement on an alkyl halide. The most significant difference is the rate enhancing effect of the alkene moiety which
has been attributed to a decrease in the activation energy of the
reaction ($\underline{9}$). A second possible mode of reaction is available
with allylic halides. This mode of displacement is usually called
S_N2' and, in general, will be promoted relative to the normal
displacement when there are substituents on the alpha carbon
which tend to inhibit the normal S_N2 pathway by inductive or
steric effects (Reaction VII).

$$RO^{\ominus}Na^{\oplus} + \; \overset{H}{\underset{R'}{\diagup\!\!\diagdown}}\!\!\overset{CH_2-X}{\diagdown\!\!\diagup} \; \longrightarrow \; RO-\underset{R'}{\overset{|}{CH}}\!\!-\!\!\overset{CH_2}{\diagup} \; + \; NaX \tag{VII}$$

   Thus, while alpha-methylallyl chloride yields only a small
amount of "abnormal" product with ethoxide ion in ethanol ($\underline{10}$),

tert-butylallyl chloride forms substantial amounts of "abnormal"
product derived through the S_N2' pathway ($\underline{11}$).
  The high electron density in the double bond system of
ethylenes makes nucleophilic attack unfavorable unless the
system is substituted with one or more electron withdrawing
groups such as $-NO_2$, $-CN$, $-COR$. When these substituents are
present, attack by alcohols or alkoxide ions occurs at the
beta-carbon predominantly. For example, researchers have
found ($\underline{12}$) that sodium methoxide or sodium ethoxide added
rapidly at room temperature to beta-nitrostyrene leads to the
alkoxide formation of the derivative (Reaction VIII). This
reaction is generally not only for arylnitroalkenes ($\underline{13}$) but
also for other activated double bonds ($\underline{14}$). Another example
of alcohol addition to an activated double bond includes the
reaction of alcohols with acrylonitrile to produce a cyano-
ethylated ether ($\underline{14A}$).

$$C_6H_5CH=CHNO_2 + RO^{\ominus} \longrightarrow C_6H_5\underset{\underset{OR}{|}}{C}H-CH_2-NO_2 \qquad (VIII)$$

  If there is a labile substituent directly attached to the
double bond, a substitution reaction can occur after the initial
attack by the anion (Reaction IX). There are several mechanisms

$$\text{(IX)}$$

which can account for the product formed, the most important of
these is the "elimination-addition" mechanism in which an intermediate acetylene is initially formed which then adds an alcohol.
  In general, the "addition-elimination" mechanism will be
favored for compounds with low electron density on the beta-carbon,
while the "elimination-addition" pathway will be favored by cis-
isomers where there is a good chance of eliminating the elements
of HX from a trans-position.

  <u>Addition to Acetylenes</u>. Since triple bonds are more susceptible to nucleophilic attack than double bonds, it might be
expected that bases would catalyze additions particularly well.
This is the case, and vinyl ethers as well as acetals may be
produced by the reaction of acetylenes with alcohols ($\underline{15}$,$\underline{15A}$)
(Reaction X).

$$R-C\equiv C-R + ROH \longrightarrow \,C=C\, \longrightarrow -\overset{\overset{OR}{|}}{\underset{|}{C}}-\overset{\overset{OR}{|}}{\underset{|}{C}}-OR \qquad (X)$$

  As with alkenes, the addition of alcohols to acetylenes is
facilitated by the presence of electron-withdrawing substituents

and is believed to proceed by a trans-mechanism($\underline{16}$). Thus,
addition of alcohols to an acetylene dicarboxylate gives mainly
the trans-compound (Reaction XI); some exceptions occur at high
temperatures and with dicyanoacetylenes where the cis-isomer is
sometimes produced.

$$ROOC-C{\equiv}C-COOR + R'OH \longrightarrow \quad \underset{R'O \qquad COOR}{\overset{ROOC \qquad H}{C=C}} \qquad (XI)$$

Because of the high electron density of the aromatic systems,
nucleophilic aromatic substitutions usually occur only where the
ring is substituted with one or more electron-withdrawing groups
ortho and/or para to the position of substitution.

In general, the reactions are second-order, first-order with
respect to both nucleophile and substrate ($\underline{17}$). The relative
activating effects of various substituents have been determined
($\underline{18}$) to be in the order:

$$-N_2^{\oplus} > -NO_2 > -SO_2CH_3 > -\overset{\oplus}{N}-(CH_3)_3 > -CN > CF_3$$

<u>Esterification</u>. Without a doubt, the best known nucleo-
philic reaction of alcohols is the reaction with organic acids
and some derivatives, like acid anhydrides and acid chlorides,
to form esters (Reaction XII).

$$RCO_2H + R'OH \xrightarrow{\overset{\oplus}{H}} RCO_2R' + H_2O \qquad (XII)$$

Esterification is an acid catalyzed reversible reaction which
is known to proceed according to the following mechanism:

$$\underset{RC-OH}{\overset{O}{\parallel}} + H^{\oplus} \rightleftharpoons \underset{R-C-OH}{\overset{\overset{\oplus}{OH}}{\parallel}} \xrightarrow{R'OH} \underset{\underset{OH}{\overset{|\,\dot{H}}{R-C-O\overset{\oplus}{-}R'}}}{\overset{OH}{|}} \rightleftharpoons \underset{\underset{OH}{R-C-OR'}}{\overset{\overset{\oplus}{OH}_2}{|}} \longrightarrow$$

$$\underset{R-C-OR'}{\overset{O}{\parallel}} + H_2O + H^{\oplus}$$

The best catalysts for this reaction are inorganic acids
(H_2SO_4, HCl), organic acids (p-toluenesulfonic, methanesulfonic)
and metal compounds such as tin and titanium derivatives - e.g.
tetraisopropyl titanate. To achieve good yields of products,
not only is a catalyst generally necessary but also the means to
drive the equilibrium to the right as written in the reactions
above.

There are many ways of changing the equilibrium, among which
are: (a) The addition of an excess of a reactant, (b) the removal
of the ester or more commonly the water by distillation using an
azeotroping agent, and (c) the removal of water by a dehydrating
agent. An example is the commercial preparation of ethyl acetate
from an aqueous solution of ethanol, acetic acid and sulfuric
acid. It happens that the lowest-boiling liquid is a ternary
mixture of ethyl acetate (83.2%), ethanol (9%), and water (7.8%).
In the final steps of the process the ethanol is removed by
washing with water. Many of the simpler esters can be made in
this way.

The necessity for the continuous removal of water can be
avoided by operating in a system composed of an aqueous and a non-
aqueous layer. When a mixture of adipic acid, methanol, sulfuric
acid, and ethylene chloride is heated, dimethyl adipate passes
into the ethylene chloride layer; the lower layer contains the
water (19).

Esters can also be made in satisfactory yields by heating an
alcohol with the ammonium salt of an acid under conditions per-
mitting removal of both ammonia and water from the reaction
mixture. The method is general and is especially recommended
where acid conditions are deleterious to the reactants. An
example is the synthesis of 2-ethylhexyl glycolate (20) (Reaction
XIII).

$$\text{(Reaction XIII)}$$

$$HOCH_2CO_2NH_4 \; + \; \underset{C_4H_9}{\overset{C_2H_5}{\diagdown}}CHCH_2OH \longrightarrow HOCH_2CO_2CH_2CH\underset{C_4H_9}{\overset{C_2H_5}{\diagup}} \; + \; NH_3 \; + \; H_2O$$

In general, the acid catalyzed esterification of organic
acids can be accomplished easily with primary or secondary alkyl
or aryl alcohols, but tertiary alcohols usually give carbonium
ions which lead to dehydration. The structure of the acid is
also of importance. As a rule, the more hindered the acid is
alpha to the carbonyl carbon the more difficult esterification
becomes (20A).

Even more facile than the reaction of an acid with an alcohol
is the reaction of an alcohol with an acyl halide (Reaction XIV).

$$\overset{\textstyle O}{\overset{\textstyle \|}{R}}C-X + R'OH \longrightarrow RCOOR' + HX \qquad\qquad\qquad (XIV)$$

The reaction is of very wide scope (21), and many functional
groups do not interfere. A base such as pyridine is frequently
added to combine with the HX formed. The alcohol may be primary,

secondary, or tertiary alkyl, or aryl. Enolic esters can also be
prepared by this method, although C-acylation can be a side
reaction.

When phosgene is the acyl halide, haloformic esters ($\underline{22}$)
or carbonates may be obtained (Reaction XV).

$$\overset{O}{\overset{\|}{Cl-C-Cl}} + 2ROH \longrightarrow RO\overset{O}{\overset{\|}{C}}OR + 2HCl \qquad (XV)$$

The alcoholysis of anhydrides (Reaction XVI) is similar in
scope to the reaction of alcohols with acyl halides. The reaction is catalyzed by general esterification catalysts, but
usually they are not needed unless the anhydride is unreactive
or the di-ester (such as a phthalate) is the product sought.

$$R-\overset{O}{\overset{\|}{C}}-O-\overset{O}{\overset{\|}{C}}-R + R'OH \longrightarrow RCOOR' + RCOOH \qquad (XVI)$$

<u>Reactions with Aldehydes and Ketones</u>. Alcohols may combine
additively with other carbonyl compounds; such addition compounds
are known as hemiacetals or acetals (Reaction XVII).

$$RC\overset{O}{\underset{H}{\diagdown}} + R'OH \xrightarrow{H^{\oplus}} RCH\overset{OH}{\underset{OR'}{\diagup}} \xrightarrow{R'OH} R-CH\overset{OR'}{\underset{OR'}{\diagup}} \qquad (XVII)$$

When the reaction is carried out with a ketone the product is
known as a ketal. With low molecular weight unbranched aldehydes
and ketones the equilibrium lies to the right. If it is desired
to make acetals or ketals of higher molecular weight molecules,
the removal of water is necessary to drive the equilibrium to the
right ($\underline{23}$).

Aldehydes and ketones may ·be converted to ethers by hydrogenation in an acidic alcoholic solution ($\underline{24}$) containing platinum
oxide (Reaction XVIII).

$$R-\overset{O}{\overset{\|}{C}}-R' + R''OH + H_2 \xrightarrow[H^{\oplus}]{PtO_2} R-\underset{OR''}{\overset{|}{C}H}-R' + H_2O \qquad (XVIII)$$

Addition to Other Unsaturated Molecules. When isocyanates are treated with alcohols, substituted methanes or carbamates are prepared (Reaction XIX).

$$R-N=C=O + R'OH \longrightarrow R-NHCOOR' \qquad (XIX)$$

The reaction gives good yields and is of wide scope. Cyanic acid (HNCO) gives unsubstituted carbamates. Although the oxygen of the alcohol is certainly attacking the carbon of the isocyanate, hydrogen bonding complicates the details of the mechanism and the kinetic picture (25).

In a very similar fashion, alcohols react with ketenes to give esters (26) (Reaction XX).

$$\diagdown C=C=O + R'OH \longrightarrow \diagdown CH-COOR' \qquad (XX)$$

The addition of HCl to a mixture of an alcohol and a nitrile in the absence of water leads to the hydrochloride salt of the iminoester (27) (Reaction XXI).

$$RC\equiv N + R'OH + HCl \longrightarrow R-\underset{\underset{OR'}{|}}{C}=NH_2^{\oplus} Cl^{\ominus} \qquad (XXI)$$

This reaction is known as the Pinner synthesis. The salt formed may be converted to the free imino ester by treatment with a weak base. It may also be converted to the corresponding ester by an aqueous acid catalyzed hydrolysis. The Pinner reaction is of a general nature and is applicable to aliphatic, aromatic and heterocyclic alcohols.

Alkoxylation. The reaction of alcohols with ethylene oxide gives polymeric products in which many units of the ethoxy group are incorporated (Reaction XXII). The reaction can be controlled

$$ROH + n\ \overset{O}{\underset{CH_2-CH_2}{\diagup\diagdown}} \xrightarrow{\ OH^{\ominus}\ } R(OCH_2CH_2)_n OH \qquad (XXII)$$

by varying reaction conditions. Propylene oxide undergoes the same type of reaction although not as fast due to the hindrance of the methyl group.

Finally, although the hydroxyl group of most alcohols can seldom be cleaved by hydrogenation, certain alcohols such as benzyl are susceptible and often readily undergo reduction (28). The most common catalysts are copper chromite and palladium-on-charcoal. Mixtures of $AlCl_3$ and $LiAlH_4$ have also been used for

this purpose (29). Though the mechanism of alcohol hydro-
genolysis is obscure, in some cases nucleophilic substitution
has been demonstrated (30).

<u>Displacement of the Hydroxyl Group</u>

Alcohols are among the most easily obtained reagents of
organic chemistry. Because of this, the overall conversion of
ROH ⟶ RX is of great importance where X is typically a halide,
hydride, azide, alkyl or an amine.
This section will provide a survey of the importance in
syntheses of replacing a hydroxyl group by other functional
groups.

<u>Alkyl Halides</u>. The classical method for converting alcohols
to alkyl iodides (31) involves heating the alcohol with iodine in
the presence of phosphorus (Reaction XXIII). Like other iodi-

$$6ROH + 2P + 3I_2 \longrightarrow 6RI + 2H_3PO_3 \qquad\qquad (XXIII)$$

nations using these reagents, the reaction proceeds through an
intermediate ester which is decomposed by the <u>in situ</u> generated
hydriodic acid.
For the preparation of bromides and chlorides from alcohols,
the corresponding acids, HBr and HCl, are the reagents of
choice (32). The mechanism (33) for this reaction is believed
to involve a protonated intermediate (Reaction XXIV) which is
further attacked by the halide.

$$ROH + HX \longrightarrow RO\overset{\oplus}{H}_2 \xrightarrow{\ X^{\ominus}\ } RX + H_2O \qquad\qquad (XXIV)$$

The observed reactivity gradation for this type of reaction
is for the acid: HI > HBr > HCl > HF, and for the alcohols
tertiary > secondary > primary.
Other halogenating agents include Ph_2PCl_3 (34), PBr_5 (35),
AlI_3 (36), and many sulfur containing reagents (37) of which
only thionyl bromide and thionyl chloride have attained wide
application. The ultimate choice of the halogenating agent to
be used will depend on the stereochemistry desired of the final
product (38,39).

<u>Amination</u>. Very few reactions of general scope exist for
the direct conversion of alcohols to amines. Among one of the
oldest is the Bucherer reaction which is used to convert
naphthols (40) and phenols (41) to their amine derivative by
reaction with aqueous sodium bisulphite and ammonia (Reaction
XXV).

$$ROH + NaHSO_3 \xrightarrow{\quad NH_3 \quad} RNH_2 + NaHSO_3 + H_2O \qquad (XXV)$$

The Ritter reaction (42,42A) is a general method for converting alcohols to amines by reaction with a nitrile and a
strong acid (Reaction XXVI). In this reaction only tertiary,

$$ROH + R'CN \xrightarrow{\;H^{\oplus}\;} RNHCOR' \xrightarrow{\;H^{\oplus}\;} RNH_2 + RCO_2H \qquad (XXVI)$$

secondary and benzylic alcohols react as they form the most
stable carbenium ions.
    Of commercial interest is the direct reaction of alcohols
with ammonia at elevated pressures and temperatures in the
presence of a dehydrating catalyst such as alumina gel. This
process is known as ammonolysis and gives a mixture of primary,
secondary and tertiary amines (Reaction XXVII).

$$ROH + NH_3 \longrightarrow RNH_2 + R_2NH + R_3N \qquad (XXVII)$$

Direct displacements of the hydroxyl group by azide are
uncommon, but carbonium ions derived from alcohols are attacked
by the azide ion to give organic azides (43) (Reaction XXVIII).

$$R_3COH + NaN_3 \xrightarrow{\;H^{\oplus}\;} R_3CN_3 + NaCl \qquad (XXVIII)$$

These azides can be further reacted (44) to give an amine as the
final product.

    <u>Tosylates</u>. Even though the S_N2 reaction cannot be performed
on alcohols, the hydroxyl group can be transformed to a good
leaving group by its reaction with p-toluenesulfonyl chloride
($p-CH_3C_6H_4SO_2Cl$). Such a group is then easily displaced by a
variety of nucleophiles in essence achieving the substitution
of the hydroxyl group (44A). By far this is one of the most
useful methods for converting an alcohol to an alkane, an ester,
an amine or any other derivative of a nucleophile.

<u>Dehydration of Alcohols</u>

    The dehydration of alcohols is an example of a wide range
of elimination reactions having the following general form
(Reaction XXIX).

$$RCH_2-CRHOH \longrightarrow RCH=CRH + H_2O \qquad (XXIX)$$

The dehydration of alcohols can take place both in solution (45) and in the gas phase (46). The general rules for this type of elimination have also been recently reviewed (47). For this reason, it will not be attempted in this section to fully explore the area; instead, only a brief review will be given.

In aqueous acidic solutions of either Brønsted or Lewis acids, the dehydration of alcohols leads to the formation of Saytzeff olefins (48). Dehydration occurs most readily if the alcohol is tertiary. For example, the formation of 1,1-diphenylethylene from methyldiphenyl carbinol (Reaction XXX)

$$(C_6H_5)_2-\overset{\overset{\displaystyle OH}{|}}{C}-CH_3 \quad \xrightarrow{\ H_2SO_4\ } \quad (C_6H_5)_2C=CH_2 + H_2O \qquad (XXX)$$

occurs very rapidly just by heating the alcohol with dilute sulfuric acid.

Isoprene has been prepared by dehydration of 3-methyl-1-butene-3-ol (Reaction XXXI) and butadiene from the dehydration

$$(CH_3)_2\ \overset{\overset{\displaystyle OH}{|}}{C}-CH=CH_2 \longrightarrow CH_2=\overset{\overset{\displaystyle }{\underset{\underset{\displaystyle CH_3}{|}}{C}}}{}-CH=CH_2 + H_2O \qquad (XXXI)$$

of 1,3-butanediol (Reaction XXXII).

$$CH_3-\overset{\overset{\displaystyle OH}{|}}{CH}-CH_2-CH_2-OH \longrightarrow CH_2=CH-CH=CH_2 + 2H_2O \qquad (XXXII)$$

The dehydration of 2-pentanol by the sulfuric acid method is of interest since it illustrates the rule that beta-olefins formation is thermodynamically the favored pathway (Reaction XXXIII).

$$CH_3CH_2CH_2-\overset{\overset{\displaystyle OH}{|}}{CH}-CH_3 \quad \xrightarrow{\ H_2SO_4\ } \quad CH_3CH_2CH=CH-CH_3 + H_2O \qquad (XXXIII)$$

Alcohols in which the beta-hydrogen is activated by a double bond undergo dehydration by concentrated alkaline media to produce dienes (49) (Reaction XXXIV). Other products such as ethers are possible when the reaction is done in the presence of dimethyl sulfoxide (50).

$$\xrightarrow{\ OH^{\ominus}\ } \qquad + H_2O \qquad\qquad (XXXIV)$$

When some straight and branched-chain aliphatic alcohols,
such as n-propanol, n-butanol and isopropanol, are subjected to
high temperatures, dehydrogenation products predominate over
dehydration (51). Presumably the eliminations take place via a
six-membered transition state and are catalyzed by hydrogen
halides in the homogeneous phase (52) to produce olefins. On
the other hand, gas phase dehydration over solid catalysts is
a valuable process for the preparation of olefins and ethers.

The most studied dehydration catalysts are the metal oxides
(53), but the selectivity of these catalysts in terms of dehydration versus dehydrogenation is not fully understood (54).

Oxidation of Alcohols
____

Among the many agents available for the oxidation of organic
compounds, the ones most commonly used are derivatives of hexa-
valent chromium (Cr-VI) or heptavalent manganese (Mn-VII).
Chromium trioxide (CrO_3) and sodium dichromate ($Na_2Cr_2O_7$) are
converted to chromium (III) in the course of oxidations for a
total transfer of three electrons to each metal atom. With
permanganate in acidic media the manganese (II) ion is formed
for a total transfer of five electrons; in neutral or basic
media, manganese dioxide is formed with a corresponding transfer
of three electrons.

The oxidation of a secondary alcohol to a ketone is usually
accomplished with a solution of the alcohol and aqueous acidic
chromic acid in either acetone or acetic acid, with a solution of
sodium dichromate in acetic acid, or by reaction of the alcohol
with aqueous acidic chromic acid as a heterogeneous system. An
example is the oxidation of the substituted cyclohexanol below
(Reaction XXXV) with sodium dichromate in sulfuric acid (55).

The course of this dichromate oxidation can be followed spectro-
photometrically as the yellow-orange absorption at 350 nm of the
chromium VI is converted to green absorption of the chromium
III ion (56).

The probable mechanism of oxidation of alcohols by chromium
(VI) species involves the formation of chromate esters and their
subsequent decomposition to ketones (57). As a rule, in the
absence of competing side reactions, the more hindered alcohols
react faster than the less hindered compounds. It has also been

found that electron donating substituents accelerate the rate
of oxidation (58).

The previously described conditions using dichromate and
chromic acid are sufficiently vigorous to slowly oxidize other
reactive centers in the molecule such as ethers, amines, carbon-
carbon double bonds, and benzylic and allylic C–H bonds. To
prevent this, a milder method of oxidation can be used, consisting of adding an aqueous solution of chromic acid (Jones
reagent) to an acetone solution of the alcohol to be oxidized
(59). Another reagent as mild as Jones reagent consists of a
chromium trioxide pyridine complex. This compound can be used
for the oxidation of alcohols containing acid sensitive functions
such as acetals and ketals (60). A convenient and inexpensive
procedure for the oxidation of secondary alcohols to ketones has
been reported (60A) to involve reaction of the alcohol with
sodium hypochlorite in acetic acid. The yield of the corresponding ketone is around 90 percent with many alcohols.
Primary alcohols react slowly giving dimeric ester, presumably
via hemiacetal intermediates.

Tertiary alcohols are relatively inert to oxidation by
chromic acid; however, tertiary 1,2-diols are rapidly cleaved by
chromic acid provided they are capable of forming a cyclic chromate ester (61) (Reaction XXXVI).

$$
\begin{array}{ccc}
\underset{\substack{|\\CH_3}}{\overset{CH_3}{\bigcirc}}\text{--OH} & \xrightarrow[\;H_2O\;]{Na_2Cr_2O_7\;\;HClO_4} & \text{(XXXVI)}
\end{array}
$$

Monohydric alcohols react rapidly with lead tetraacetate to
form alkoxy lead (IV) intermediates. These intermediates decom-
pose thermally or photolytically in a variety of ways to produce
ketones, esters, and cyclic ethers (62), as can be seen below
(Reaction XXXVII).

$$
R_1(CH_2)_4\text{--}\underset{\substack{|\\OH}}{CH}\text{--}R_2 \xrightarrow{Pb(OCOCH_3)_4} R_1\text{--}(CH_2)_4\text{--}\underset{\substack{|\\O\text{--}Pb(OCOCH_3)_3}}{CH}\text{--}R_2 \qquad \text{(XXXVII)}
$$

$$
R_1CH_2\text{-}\underset{O}{\bigcirc}\text{-}R_2 \;\text{ or }\; R_1\text{--}(CH_2)_4\text{--}O\overset{O}{\overset{\|}{C}}\text{--}CH_3 \;\text{ or }\; R_1\text{--}(CH_2)_4\text{--}\overset{O}{\overset{\|}{C}}\text{--}R_2
$$

Both the structural features of the alcohol and the reaction conditions used are important in determining which of the decomposition pathways is followed. If the lead alkoxide from a primary or secondary alcohol is formed in the presence of a donor solvent, such as pyridine, oxidation to an aldehyde or ketone is the primary mode of decomposition (63) (Reaction XXXVIII).

$$CH_3(CH_2)_3CH_2-OH \xrightarrow[\substack{Pyridine \\ 25°C}]{Pb(OCOCH_3)_4} \begin{array}{l} CH_3(CH_2)_3CHO \\ CH_3(CH_2)_3CH_2OCOCH_3 \end{array} \quad (XXXVIII)$$

When the oxidations are carried out in the presence of calcium carbonate and lead tetraacetate, the major product is cyclic ethers formed through the intermediacy of free radicals (64).

Analytical Determination of the Hydroxyl Group

Some simple "test tube" reactions can be used to determine the presence and/or the type of hydroxyl groups in organic molecules. Although modern spectroscopy has made the knowledge of this type of analytical chemistry less imperative today, this section will try to cover a few of the most important reactions which are still useful for the fast determination of the hydroxyl group.

(a) Primary and secondary alcohols will react with Nessler reagent (65), a mixture of equal volumes of 2N NaOH and potassium mercury (II) iodide solution. Mix a few drops of the substance with 5 mL of the reagent and boil the mixture for a few minutes. The presence of a primary or secondary alcohol is detected by the formation of a brownish yellow to gray precipitate which turns gray on standing. Tertiary alcohols do not react with Nessler's reagent. A white precipitate may be formed on mixing the reagents, but it dissolves on shaking the test tube.

(b) When added to a solution of n-bromosuccinimide (66), primary alcohols give a stable color, secondary alcohols a fleeting color and tertiary alcohols no color.

(c) Tertiary alcohols are dehydrated when boiled with Deniges' reagent (67); alkenes, which give a yellow precipitate, are formed. The reaction is thus of alkenes, not of tertiary alcohols. Primary alcohols do not react, but some secondary alcohols react almost as readily as the tertiary alcohols.

For a complete treatment of simple qualitative and quantitative reactions for alcohols, the reader is referred to Veibel's book (69) which deals in depth with a large number of analytical reactions.

Conclusion

In the future, there is no doubt that alcohols will play a major role not only as fuel components (70,71) but also as feed stocks for the syntheses of more complicated organic compounds (72). A great amount of research effort is presently directed to the economic conversion of CO and H_2 to methanol (73), and on homologation of methanol to higher alcohols (74). Conversion of synthesis gas from coal to acetic anhydride (75) through the intermediacy of methyl acetate (from methanol) will soon be a commercial reality.

Dramatic entries have been made into the technology of the eighties by direct utilization of CO and H_2 to produce, for example, methanol and polyhydric compounds (76) (Reaction XXXIX). This unique reaction has attracted considerable attention due to a

$$CO + H_2 \xrightarrow[\text{Bases}]{\text{Rh Catalyst}} CH_3OH + HOCH_2(CHOH)_x CH_2OH \qquad \text{(XXXIX)}$$

$$X = 0,1,2$$

relatively high specificity to ethylene glycol, and to the unusual type of catalysts which bears superficial relationship to homogeneous catalytic systems (77).

Other advances include the commercialization of a process to make oxalic acid (78) (Reaction XL).

$$2\ C_4H_9OH + 2CO + 1/2\ O_2 \xrightarrow[\text{Cat.}]{\text{Pd}} \begin{array}{c} COOC_4H_9 \\ | \\ COOC_4H_9 \end{array} \xrightarrow{H_2O} \begin{array}{c} CO_2H \\ | \\ CO_2H \end{array} \qquad \text{(XL)}$$

It is interesting that hydrogenation of the intermediate dibutyl oxalate would produce the glycol and the starting butanol.

This new area of chemistry is still in its infancy when compared to the other reactions covered in this chapter, but many exciting developments can be anticipated in the future.

Literature Cited

1. Ballinger, P. and Long, F. A., J. Am. Chem. Soc., 81, 1050 (1959).

2. Ballinger, P. and Long, F. A., J. Am. Chem. Soc., 82, 795 (1960).

3. Long, F. A. and Ballinger, P., Electrolytes (Ed. B. Pesce), Pergamon, Oxford, 1962, p. 152.

4. Taft, R. W., J. Am. Chem. Soc., 74, 3120 (1952).

5. Taft, R. W., J. Am. Chem. Soc., 75, 4231 (1953).

6. Brauman, J.I. and Blair, L. K., J. Am. Chem. Soc., 92, 5986

6A. Edwards and Pearson, J. Am. Chem. Soc., 84, 26 (1962).

7. Bateman, L. C.; Cooper, K. A.; Hughes, E. D.; and Ingold, C. K., J. Chem. Soc., 925 (1940).

7A. Grady, G. L. and Chokski, S.K., Synthesis, 483 (1972).

8. Swain, Ketley, and Bader, J. Am. Chem. Soc., 81, 2353 (1959).

9. Vernon, C.A., J. Chem. Soc., 4462 (1954).

10. Hughes, E. D., Trans. Faraday Soc., 34, 185 (1938).

11. delaMare, P.B.D.; Hughes, E. D.; Merriman, P. C.; Pichat, L. C. and Vernon, C. A., J. Chem. Soc., 2563 (1958)

12. Meisenheimer, J. and Heim, F., Ber., 38, 467 (1905).

13. Thiele, J. and Haechkel, S., Liebigs Ann Chem., 325, 8 (1902).

14. Patai, S. and Rappoport, Z., The Chemistry of the Alkenes, (Ed. S. Patai), Interscience, London, 1964, Chapter 8.

14A. Brunson, Organic Reactions, 5, 79 (1949).

15. For a review see Shostakovskii, Bogdanova, and Plotnikova, Russ. Chem. Rev., 33, 66-77 (1964).

15A. Linn, W. S., Waters, W. L. and Caserio, M. C., J. Am. Chem. Soc., 92, 4018 (1970).

16. Winterfeldt, E., Angew. Chem., Intern. Ed. Engl., B, 6, 423 (1967).

17. Bunnett, J. F. and Zahler, R. E., Chem. Rev., 49, 273 (1951).

18. Bunnett, F. J., Quart. Rev. (London), 12, 1 (1958).

19. Clinton, R. O. and Laskowski, S. C., J. Am. Chem. Soc., 70, 3135 (1948).

20. Filachione, E. M., Costello, E. J. and Fisher, C. H., J. Am. Chem. Soc., 73, 5265 (1951).

20A. Pfeffer, P. E., et al., Tetrahedron Letters, 4063 (1972).

21. Sonntag, Chem. Rev., 52, 237 (1953).

22. Matzner, Kurkjy, and Cotter, Chem. Rev., 74, 645 (1964).

23. Long and Paul, Chem. Rev., 57, 935 (1957).

24. Verzele, Acke and Anteunis, J. Chem. Soc., 5598 (1963).

25. Robertson and Stutchbury, J. Chem. Soc., 4000 (1964).

26. Lacey, Angew. Chem., 68, 361 (1956).

27. Zil'berman, Russ. Chem. Rev., 31, 615 (1962).

28. Hartung and Simonoff, Org. Reactions, 7, 263 (1953).

29. Blackwell and Hickinbottom, J. Chem. Soc., 1405 (1961).

30. Dar'eva and Miklukhin, J. Gen. Chem. USSR, 29, 620 (1959).

31. Organic Synthesis, Coll. Vol. 2, John Wiley and Sons Inc., New York, 1943, p. 322.

32. Fieser, L. F. and Fieser, M., Reagents for Organic Synthesis, John Wiley and Sons Inc., New York, 1967, p. 296.

33. Gerrard, W. and Hudson, H. R., J. Chem. Soc., 2310 (1964).

34. Hoffman, H., Horner, L., Wippel, H. G., and Michael, D., Chem. Ber., 95, 523 (1962).

35. Eliel, E. L. and Haber, R. G., J. Org. Chem., 24, 143 (1959).

36. Broome, J., Brown, B. R., and Summers, G.H.R., J. Chem. Soc., 2071 (1957).

37. Stroh, R. and Hahn, W., Houben-Weyl, Methoden der Organischen Chemie, Vol. 5, Pt. 3, Georg Thiemen, Verlag, Stuttgart, 1962, p. 857.

38. Cremlyn, R. J. and Shoppee, C. W., J. Chem. Soc., 3794 (1954).

39. Shoppe, C. W., Lack, R. E., Sharma, S. C. and Smith, L. R., J. Chem. Soc., (C), 1155 (1967).

40. White, E. H. and Woodcock, D. J., Chemistry of The Amino Group (Ed. S. Patai), Interscience, New York, 1968, p. 486.

41. Drake, N. L., Organic Reactions, Vol. 1, John Wiley and Sons, Inc., New York, 1942, p. 105.

42. Ritter, J. J. and Kalish, J., J. Am. Chem. Soc., 70, 4048 (1948).

42A. Zil'berman, Russ. Chem. Rev., 29, 331 (1960).

43. Arcus, C. L. and Mesley, R. J., Chem. Ind. (London), 701 (1951).

44. Moore, A. T. and Rydon, H. N., Org. Synth., 45, 47 (1965).

44A. Wentworth, S. E., and Sciaraffa, P. L., Org. Prep. Proc., 1, 225 (1969).

45. Saunders, W. H., The Chemistry of Alkenes (Ed. S. Patai), Interscience Publishers, London, New York, Sydney, 1964, p. 149.

46. Maccoll, A., The Chemistry of Alkenes (Ed. S. Patai), Interscience Publishers, London, New York, Sydney, 1964, p. 203.

47. Banthorpe, D. V., Elimination Reactions, Elsevier Publishing Co., Amsterdam, London, New York, 1963.

48. Grimaud, J. and Laurent, A., Bull. Soc. Chim. France, 3599 (1967).

49. Kitchen, L. J., J. Am. Chem. Soc., 73, 2368 (1951).

50. Trynelis, V. J., Hergenrother, W. L., Hanson, H. T. and Valicenti, J. A., J. Org. Chem., 29, 123 (1964).

51. Barnard, J. A., Trans. Faraday Soc., 53, 1423 (1957).

52. Failes, R. L. and Stimson, V. R., J. Chem. Soc., 653 (1962).

53. Winfield, M. E. and Emmett, P. H., Catalysis, Vol. 7, Reinhold Publishing Corp., New York, 1960.

54. Schwab, G. M. and Schwab-Agallidis, E., J. Am. Chem. Soc., 71, 1806 (1949).

55. Hussey, A. S. and Baker, R. H., J. Org. Chem., 25, 1434 (1960).
56. Westheimer, F. H. and Nicolaides, N., J. Am. Chem. Soc., 71, 25 (1949).
57. Wiberg, K. B. and Schafer, H., J. Am. Chem. Soc., 91, 927, 933 (1969).
58. Kwart, H. and Francis, P. S., J. Am. Chem. Soc., 77, 4907 (1955).
59. Eisenbraun, E. J., Org. Synth., 42, 79 (1962).
60. Holum, J. R., J. Org. Chem., 26, 4814 (1961).
60A. Stevens, Chapman and Weller, J. Org. Chem., 45, 2070 (1980).
61. Heindel, N. D., Hanrahan, E. S., and Sinkovitz, R. J., J. Org. Chem., 31, 2019 (1966).
62. Heusler, K. and Kalvoda, J., Angew. Chem., Intern. Ed. Engl., 3, 525 (1964).
63. Partch, R. E., J. Org. Chem., 30, 2498 (1965).
64. Cope, A. C., McKervey, M. A., Weinshenker, N. M., and Kinnel, R. B., J. Org. Chem., 35, 2918 (1970).
65. Rosenthaler, L., Angew. Chem., 20, 412 (1907).
66. Chem. Abs., 67, Ref. 17618 (1967).
67. Deniges, G., Compt. Rend. Acad. Sci., 126, 1145 (1898).
68. Liebermann, C., Ber., 7, 247; 806; 1098 (1874).
69. Veible, S., The Determination of Hydroxyl Groups, Academic Press, London and New York, 1972.
70. Reed, T. B., "Methanol for Fuel: A Bibliograpny on the Production and Use of Alcohol as Fuel." M.I.T. Energy Lab, Methanol Div., (July 1, 1974).
71. Mrstik, A. V., "Experience with Tert-Butyl Alcohol for Gasoline," Pet. Inf. Ed. Int., N 1505, 65.57 (7/19/79).
72. Novotny, M. and Mador, I. L., "Synthesis of C_2-Oxygenated Chemicals from Methanol," Eighth Conference on Catalysis in Organic Syntheses, New Orleans, Louisiana, June 2-4, 1980.
73. Harris, W. D. and Davison, R. R., Oil and Gas Journal, 71, 70 (1973).
74. Chen, M. J. and Feder, H. M., "Mechanism of a New Process for Methanol Homologation," Eighth Conference on Catalysis in Organic Syntheses, New Orleans, Louisiana, June 2-4, 1980.
75. Abelson, P. H., Science, 207, 479 (1980).
76. Pruett, R. L., Ann. N.Y. Acad. Sci., 295, 239 (1977).
77. Wender, I., Conference on Chemical Research Applied to World Needs, Toronto, Canada, July 10-13, 1978.
78. Technocrat, 11 (4), 78 (1978).

RECEIVED March 13, 1981.

Methanol: Manufacture and Uses

THOMAS F. KENNEDY and DEBORAH SHANKS

Celanese Chemical Co., Inc., Dallas, TX 75247

This paper covers the current technology of methanol production, reviews how the energy crisis and the escalation of hydrocarbon feedstocks impact that technology, and describes conventional, new, and potential uses for methanol. Methanol is a chemical intermediate and solvent produced from several feedstocks and is consumed in a variety of end uses. Prior to the development of a synthetic route to methanol, commercial quantities were obtained from the destructive distillation of wood or other biomass. Now, with interest focused on conservation and the use of renewable resources, methods have been proposed to use biomass again as a methanol feedstock. While some of these proposals hold great appeal, they are still speculative and beyond the scope of this paper. (Methanol from wood is discussed in the following chapter.)

<u>Methanol Production</u>

Methanol is produced by the catalytic reaction of carbon monoxide and hydrogen (or synthesis gas) in a converter according to the reaction:

$$CO + 2H_2 \longrightarrow CH_3OH$$

Figure 1 depicts a typical methanol synthesis scheme:

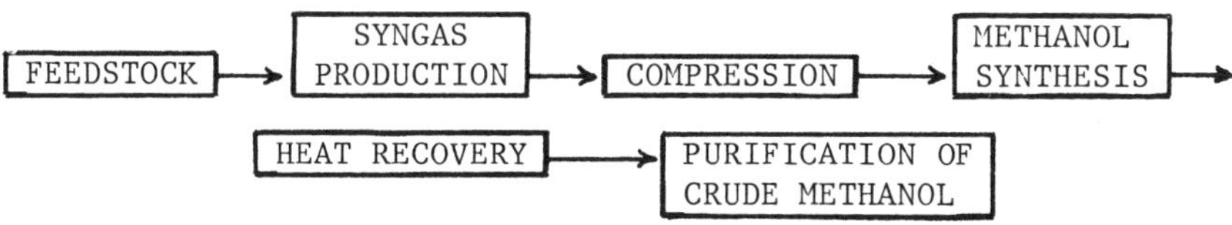

Figure 1.   Methanol synthesis summary

<u>Synthesis Gas</u>. There are three principal routes to synthesis gas: steam reforming, partial oxidation, and coal gasifica-

tion. Steam reforming is limited to lighter hydrocarbon feedstocks, primarily methane and naphtha. Partial oxidation can use
not only those light feedstocks, but also heavier feedstocks like
residual oil. Coal gasification, applicable to all types of
coal, anthracitic to lignitic, is an emerging technology with
several competing processes (1).

Thus a variety of hydrocarbons, ranging from natural gas to
coal, are used in methanol production. Regardless of the
feedstock used to prepare the synthesis gas, it is necessary to
remove sulfur so that the converter catalyst is not poisoned.
Before natural gas or naphtha is reformed, the feedstock is
desulfurized. In the partial oxidation and coal gasification
processes, the feedstock is first oxidized and the resulting
synthesis gas is desulfurized before entering the converter.

Methanol Synthesis. By whatever means produced, synthesis
gas is then compressed for feed to the converter. There are two
routes for methanol synthesis, a high pressure process requiring
compression to about 340 atmospheres, and a low pressure process
requiring compression in the range of 50 to 100 atmospheres (2,2A).
Because of the inherent economic advantages of the low pressure
technology, high pressure units are in the process of being
phased out or converted to low pressure. ICI and Lurgi have the
two principal low pressure technologies practiced today for
methanol synthesis. They differ primarily in the technique used
to remove the heat of reaction and thus control temperatures
within the converter. This difference results from converter
design.

Converter exit gas containing methanol is cooled by heat
exchange with cooling water. The condensed methanol and water
mixture is then separated. This crude methanol is purified in a
two or three column distillation. The first column separates
light ends from methanol. The second column separates methanol
from water and fusel oils. If very low ethanol content is
required, a third column can be used for final purification.

Impact of Energy Crisis. The two principal steps in methanol production, synthesis gas preparation and methanol synthesis,
have been greatly impacted by the continuing energy crisis.
Synthesis gas production has been influenced by the relative
value of its hydrocarbon feedstocks, and methanol synthesis by
the improvements in the energy balances of the newer low pressure
technologies. Most synthesis gas production for methanol is
based on steam reforming of natural gas. Until recently, many
units outside the U.S. utilized naphtha as a feedstock. Steam
reforming of naphtha produces a synthesis gas containing
hydrogen and carbon in a ratio close to theoretical requirements
for methanol synthesis. As the price of crude oil has
skyrocketed, the price of naphtha has followed and made naphtha
uneconomic versus natural gas as a methanol feed. In the U.S.
natural gas has always been the economic choice because of the

relative abundance of reserves that existed on the Gulf Coast.
While the energy crisis has caused a tremendous increase in the
price of natural gas, the even more dramatic increase in world
oil prices makes natural gas the economic choice for feedstock
over naphtha. However, steam reforming of natural gas produces
a hydrogen to carbon oxides ratio of about 3:1. To satisfy the
stoichiometric requirements for the methanol synthesis reaction,
an external source of carbon, usually carbon dioxide, can be
added to the feed, and/or excess hydrogen can be purged from
unreacted synthesis gas and used as a fuel for the reformer.

Partial oxidation has more recently attracted attention
because of its ability to utilize the least valuable portion of
the crude oil barrel (3). Partial oxidation of residual oil
generates synthesis gas with a hydrogen to carbon oxides ratio of
about 1:1. To adjust the synthesis gas to the required
composition, a portion of the gas stream is sent to a shift
converter where CO and water are converted to hydrogen and CO_2
according to the water gas shift reaction:

$$CO + H_2O \longrightarrow CO_2 + H_2$$

The carbon dioxide is removed before returning the hydrogen to
the make up gas stream.

The escalation of oil prices has caused even residual oil
prices to rise to a point where coal has attracted interest as a
feedstock for synthesis gas in methanol production. Several
plants now exist in other parts of the world based on coal
gasification. Coal is gasified in the presence of oxygen and
steam at high temperatures. Like synthesis gas prepared from
residual oil, the hydrogen to carbon oxides ratio is about 1:1
and must be adjusted to a higher ratio utilizing the water gas
shift reaction. Table I summarizes the feedstock, process, and
hydrogen:carbon oxides ratio of the competing feedstocks.

TABLE I

METHANOL FEEDSTOCKS

	FEEDSTOCK	SYNGAS MANUFACTURE PROCESS	HYDROGEN:CARBON OXIDES RATIO
1.	NATURAL GAS	STEAM REFORMING	3:1
2.	NAPHTHA	STEAM REFORMING	2:1
3.	RESIDUAL FUEL OILS	PARTIAL OXIDATION	1:1
4.	COAL	GASIFICATION	1:1

The energy picture for the future remains cloudy, and the
relative costs of competing feedstocks difficult to project with

any certainty. Increasingly it appears that coal will be the
long term economic choice for synthesis gas feedstock. Yet, in
those areas of the world where natural gas is in excess supply,
gas remains the more attractive feedstock for synthesis gas in
methanol production. The capital costs are much lower. The
hydrogen:carbon oxides ratio is better for methanol synthesis, and
the sulfur removal step is minimized. Oxygen is not required, and
the solids handling problem is eliminated. For these reasons, it
is likely that methanol plants will be built outside the U.S. in
the 1980's in areas of excess natural gas supplies, and methanol
from these plants will supply part of U.S. demand.

As previously stated, the high pressure process to convert
synthesis gas to methanol is being phased out in favor of the
low pressure process because of the latter's inherent economic
advantages. These advantages include lower capital costs, lower
energy requirements and lower maintenance costs. The first
commercial synthesis of methanol had been carried out in a
reactor at a pressure of about 340 atmospheres. High pressure
processes use a zinc/chromium catalyst that is rugged and poison
resistant. However, these catalysts exhibit poor activity,
necessitating high temperatures (325°-375° C) and pressures for
commercial use (4).

Most existing units are low pressure processes or are being
converted to low pressure technology. Reduction in synthesis
pressure requires a reduction in temperature. Lower
temperatures necessitate a more active catalyst. The activity
and selectivity of copper based catalysts for methanol synthesis
had been known prior to commercial utilization (2). The
stumbling block, preventing earlier use of the catalyst, is the
sensitivity of the catalyst to sulfur. This has been solved by
the development of methods to remove sulfur from natural gas
before it is reformed (4). Since the first low pressure
methanol plant was started up in 1966 (2), most new plants have
been of that design, and the energy crisis has resulted in the
ongoing conversion of the remaining high pressure units to the
low pressure process.

Current Applications

Methanol has long been an important item of commerce,
resulting from the availability of low cost raw materials coupled
with the development and refinement of an efficient synthetic
process. Despite having been directly and greatly impacted by
the ongoing energy crisis, methanol remains a relatively inexpensive solvent and chemical intermediate with a myriad of uses.
While many of these uses are mature with only minimal growth
forecast, newer end uses continue to be commercialized, spurred
by economic factors, namely availability and low cost. Today,
over three billion gallons of methanol are produced and consumed
in the world annually with the U.S. accounting for nearly one

third of the total. Growth is forecast to approximate twice the
rate of growth in GNP (5).

Conventional uses of methanol account for 90% of present
consumption and include formaldehyde, dimethyl terephthalate,
methyl methacrylate, methyl halides, methylamines and various
solvent and other applications. Newer uses for methanol that
have revitalized its growth and outlook include a new technology
for acetic acid, single cell protein, methyl tertiary butyl
ether-(MTBE), and water denitrification. Potential uses for
methanol include its use as a carrier for coal in pipelines, as
a source of hydrogen or synthesis gas used in direct reduction
of iron ore, as a direct additive to or a feedstock for
gasoline, peak power shaving and other fuel related
possibilities. Table II lists the world methanol demand by end
use in 1979.

The largest and oldest chemical intermediate use for
methanol is formaldehyde. Over half of the methanol currently
consumed in the world goes into formaldehyde production.
Formaldehyde is produced by the catalytic oxidation or the
oxidative dehydrogenation of methanol. The major outlet for
formaldehyde is amino and phenolic resins. These resins are in
turn used in the manufacture of adhesives for wood products,
molding compounds, binders for thermal insulation and foundry
resins. Formaldehyde is also consumed in the production of
acetal resins, pentaerythritol, neopentyl glycol,
trimethylolpropane, methylenediphenyldiisocyanate (MDI), and
textile treating resins.

Dimethyl terephthalate (DMT) is produced either by the
esterification of terephthalic acid or the esterification of
monomethyl terephthalate produced by oxidation of methyl
p-toluate. DMT is consumed in the production of polyethylene
terephthalate, the polymer used in the manufacture of polyester
fibers, films and bottle resins. Terephthalic acid (TPA) is
also used in the production of polyethylene terephthalate but
does not consume methanol. Since TPA is continuing to increase
its share of the market, DMT is expected to exhibit slower growth
than the overall market for polyethylene terephthalate.

Methyl methacrylate, accounting for 4% of methanol
consumption, is produced by the cyanohydrin process utilizing
methanol. Methyl methacrylate is used to produce acrylic sheet,
surface coating resin, and molding and extrusion powder. Also,
there exist minor miscellaneous uses such as modification of
acrylic fiber and polyester resin.

Methanol consumed in the production of methyl halides and
methylamines accounts for 8% of consumption. Methyl chloride is
made by the reaction of hydrochloric acid and methanol.
Methylene chloride and chloroform are produced by chlorinating
methyl chloride. Methylamines are produced by catalytically
reacting ammonia with methanol. Methyl chloride is used in the
production of silicones and tetramethyl lead. Methylene

TABLE II

WORLD METHANOL DEMAND

1979

| | PERCENT OF TOTAL
METHANOL CONSUMPTION |
END USE	
Formaldehyde	52
Dimethyl Terephthalate	4
Methyl Methacrylate	4
Methyl Halides and Methyl Amines	8
Acetic Acid	6
MTBE	4
Single Cell Protein	<1
Solvents	8
Miscellaneous	14
	100%

chloride is used as a solvent, urethane blowing agent and
aerosol propellant. Chloroform's principal uses are in the
manufacture of propellants and in the production of flurocarbon
plastics. Monomethylamine is consumed in the production of
insecticides and surfactants. Dimethylamine is used in the
production of spinning solvents for acrylic fibers, surfactants,
insecticides and rubber chemicals.

Solvents account for 8% of methanol demand and include
process uses such as extracting, washing, drying and
crystallizing. Miscellaneous uses of methanol include the
production of glycol methyl ethers, methyl acrylate and methyl
acetate. Other uses include antifreeze, gasoline deicer,
windshield washer fluid and hydrate inhibition in natural gas.

New Applications

As seen from the above, conventional uses of methanol cover
a wide range of products which in turn find application in a very
broad cross-section of industrial and consumer goods. New end
uses have continued to develop and spur the growth of methanol
production. One such development is the Monsanto low pressure
process that carbonylates methanol to acetic acid (6). Essentially all new acetic acid capacity now being installed is based
on Monsanto technology. By 1981, eleven plants converting
methanol to acetic acid are scheduled to be on stream. At
capacity they will consume over 300 million gallons of methanol.
Another emerging use of methanol is methyl tertiary butyl
ether (MTBE), a compound which improves the octane rating of
gasoline (7). It is produced by the selective reaction of methanol with isobutylene over ion exchange resin catalyst. Government regulations requiring unleaded gasoline and improved mileage
provide impetus for the development of an octane improver which
does not add lead to the atmosphere or affect automobile
hydrocarbon emissions. MTBE, while not a replacement for
tetraethyl lead or MMT (methylcyclopentadienyl manganese
tricarbonyl), is a relatively high octane blending stock for
gasoline. MTBE has been produced in Europe since 1973. There
is capacity in Europe to produce 220 thousand tons of MTBE which
requires about 27 million gallons of methanol. In the U.S., the
EPA gave approval in early 1979 for the addition of MTBE to
unleaded gasoline. Since then several units have been brought
on stream in the U.S. Present U.S. MTBE capacity is 380
thousand tons which requires about 48 million gallons of
methanol. When all the announced plants are brought on stream
and run at capacity, 180 million gallons of methanol will be
required for MTBE worldwide.
Methanol appears to be the preferred carbon source for
single cell protein production though several hydrocarbons could
be used. Bacteria is grown in a mixture of air and methanol
containing ammonia, water and other mineral nutrients. An ICI

plant that recently was brought on stream has the capacity to
produce 75 thousand metric tons of SCP for primary use as an
animal feed additive (8). When run at capacity, the plant will
consume 50 million gallons of methanol. An optimistic potential
for methanol consumption in SCP production would be in the
several hundred million gallons range by the mid-1980's.
Whether this projection is realized or not depends on 1) the
ability of single cell protein to compete with soybean as a
protein source, 2) successful handling of regulatory,
environmental and agricultural opposition and 3) methanol not
being displaced by a cheaper carbon source.

Very small quantities of methanol are now consumed in the
denitrification of waste water. Methanol is used as a carbon
source by bacteria which convert nitrates and nitrites to
nitrogen. Methanol is an efficient carbon source and decreases
the production of byproducts. The potential requirement for a
carbon source for water denitrification is large but the use of
methanol is threatened by cheaper feedstocks. One large
denitrification plant in the U.S., for example, uses brewery
waste as a carbon source. In the future, water treatment plants
designed for water denitrification may utilize a less expensive
alternate carbon source than methanol.

Fuel uses are a potential application which would require
substantial volumes of methanol. As mentioned earlier they are
reviewed in the following chapter. A fuel related potential use
of methanol is as a replacement for water used to carry coal in
pipelines. Methanol is being considered for this use because it
would eliminate a demand for water, which is often scarce in
areas where coal is mined, and methanol could be burned as a
fuel with the coal at its destination. Methanol has also been
touted as a good feedstock for gases used in the direct
reduction of iron ore. If this use of methanol is realized, it
will not be before the mid to late 1980's. Other potential new
uses for methanol include a feedstock for ethylene and propylene
production (9) and a feedstock for gasoline production (10).

Conclusions

In conclusion, this review traces the evolution of methanol
technology as well as its applications. This dynamic industry
is being challenged by the energy crisis. In response to this
challenge, the industry is developing efficient, flexible,
evolving technologies and looks forward to a future of continued
growth.

Literature Cited

1. Kermonde, R. I.; Nicholsen, A. F.; Jones, J. E., "Methanol from Coal: Cost Projections to 1990", <u>Chem Engineering</u>, 1980, <u>87</u>(4), 111-116.
2. <u>Hydrocarbon Processing</u>, 1979, <u>59</u>(11) 191-193.
2A. Haddeland, G. E., "Methanol By High Pressure Synthesis", Report #43, SRI Process Economics, Oct. 1968, pp. 63-104.
3. "Methanol" A Global Analysis - 1977-1990", Chem Systems, Inc., May, 1979 (Partial Oxidation of Heavy Fuel Oils) pp. 260-264.
4. Pettman, M. J., <u>Chemical Age</u>, 1979, <u>119</u>, 3143.
5. <u>Chemical Marketing Reporter</u>, Feb. 11, 1980, p. 6.
6. Takaoka, S., "Acetic Acid Synthesis From Methanol and Carbon Monoxide", Report #37A, SRI Process Economics, March, 1973, pp. 83-107.
7. Stinson, S. C., <u>C&EN</u>, 1979, <u>57</u>(26), 35-36.
8. <u>Chemical Marketing Reporter</u>, Feb. 11, 1980, pp. 16-20.
9. <u>Chem. Week</u>, 1979, <u>125</u>(5), 27.
10. <u>C&EN</u>, 1979, <u>57</u>(28), 19.

RECEIVED December 29, 1980.

Methanol from Wood in Brazil

VICTOR BRECHERET, JR. and ANTONIO JOSÉ AYRES ZAGATTO

Companhia Energética De São Paulo, São Paulo, S.P., Brazil

The Brazilian economy has been severely damaged by continuous increases in crude oil prices. Domestic production of fossil fuels is small, and this year oil imports will demand 40% of all export revenues. There has been a chronic deficit in the balance of payments and monies spent are not recycled within the country. Also, other countries that traditionally import Brazilian goods tend to adopt protectionist trade policies to balance their own external trade.

The primary energy consumption in Brazil is as shown in Table I (<u>1</u>):

Table I
Primary Energy Consumption in Brazil (1977)

Source	Daily consumption	
	(1 000 Bbl)	(%)
Imported oil	720.3	34.9
Brazilian oil	141.0	6.8
Imported coal	54.0	2.6
Brazilian coal	28.1	1.4
Hydroelectricity	539.1	26.1
Wood	417.7	20.2
Others (natural gas, ethanol, charcoal, bagasse)	164.8	8.0
TOTAL	2,065.0	100.0

Imported oil in the Brazilian energy matrix does not appear large when compared with most of the OECD countries. The major problem is that Brazil has great difficulty increasing exports to compensate for the continuing rise in oil prices. Brazil is practically compelled to reduce its imports. The gross national product must increase at least 6% per year to avoid social problems that would come with unemployment. The energy demand increase associated with this growth is around 5% (historically this rate has been around 10% annually over the past 20 years). Annual energy consumption per capita is around 900 kg of oil

equivalent (20th place in the world ranking) which indicates that
conservation policies will lead to small but not negligible
effects. On the other hand, oil prices are expected to rise
until they reach the cost of producing the cheapest large scale
synthetic fuel. Under such conditions it is reasonable for
Brazil to produce synthetic fuels, even with costs higher than
the present oil price.
      The energy blend in Brazil is not very dependent on imported
oil (85% of 42% = 35.7 in 1978) (2). Therefore, the partial
substitution of a more expensive domestically produced synthetic
fuel for imported oil would have little impact on average
energy costs for the country. As oil prices tend to rise,
possibly up to the level of synthetic fuels costs, the risk of
having too expensive energy is very small in Brazil. On the
contrary, Brazil may obtain quite an attractive situation with
regard to energy supplies.

Biomass Energy in Brazil

      Brazil is a very large country with a warm and rather humid
climate, which permits large, economical production of biomass.
Nowadays, wood and charcoal are responsible for 25.7% of the
total energy supply in the country. So, the intensive utilization
of energy from biomass does not represent something really new.
In the forties more than 50% of its energy was obtained from
biomass (mainly wood). In order to produce liquid fuels from
biomass, basically two routes can be adopted:
      - production of ethanol through fermentation;
      - production of gas, followed by synthesis of fuels
(methanol for example).
      Potential production rates of energy (expressed in Mcal/
hectare/year) in the form of alcohol, through biomasses that are
already extensively planted in Brazil, are shown in Table II (3):

Table II
Potential Production Rates of Energy in Brazil

Crops	Fraction of Occupied Land	Energy (alcohol)	Agricultural Production Energy	Net Liquid Energy
Sugar cane	100 %	18,020	3,796	14,224
Cassava	81 %	10,332	2,431	7,091
Eucalyptus (ethanol)	87 %	12,373	551	11,822
Eucalyptus (methanol)	73 %	18,407	551	17,856
Pinus (ethanol)	85 %	16,464	471	15,993
Pinus (methanol)	68 %	21,362	471	20,891

      Fraction of occupied land represents the fraction of the
total area occupied by crops; the remaining part is occupied by
Eucalyptus culture, for supplying energy to the plant, if
necessary. The following raw material production per hectare
per year is assumed:

- eucalyptus: 11.8 tons (oven dried material)
- pinus: 14.6 tons (oven dried material)
- sugar cane: 55 tons

Brazil has produced ethanol as a fuel since 1930. Table II shows that productivity of methanol from wood is better than ethanol from sugar cane and much better than ethanol from wood. This fact plus the convenience of not being dependent on just one crop and the possibility of using poor lands for reforesting indicated that the production of methanol from wood should be more deeply evaluated.

The Methanol Option

Brazil made an option for alcohol fuels in 1975, when it started the PNA - Programa Nacional do Álcool - that led to the production of nearly four million cubic meters of ethanol in 1979, equivalent to 20% of the total gasoline consumption. The necessary investment was small, as the international market for sugar had undergone a very severe crisis, with very low prices and little demand, between 1974 and 1978. Investments made in sugar cane production and sugar mills, that would otherwise remain useless, have been utilized to produce ethanol for marginal costs. The whole agricultural, industrial and managerial infrastructure was available, and so ethanol production grew at a 35% yearly rate between 1970 and 1978. It is expected to grow at 17% yearly for the next eight years, reaching ten million cubic meters in 1986.

Even considering the very unique conditions that Brazil has to produce ethanol, its cost is quite high (around US\$ 1.10/ gallon). Furthermore, as said before, it is not desirable for Brazil to remain so dependent on sugar cane. Methanol from wood appears to be the natural complement to the existing ethanol industry and market. The cost of methanol production seems very attractive. Its large scale production does not present any serious technological problem. Only the gasification system to process large amounts of wood has to be developed, but this is not a difficult problem. Brazil has considerable experience in reforestation, for cellulose and charcoal production. More than three million hectares are already planted and 300,000 hectares are being planted yearly. Increasing this rate of reforesting is not a major problem. The most important point is to prove the economic feasibility of methanol production in Brazil.

Production Routes

Figures 1 and 2 show the process routes for methanol synthesis. The first one is based on a fluidized bed and includes a pressurized gasifier (Winkler type) that produces a very clean gas, practically free of tars and other impurities from wood distillation. The second route is based on a fixed bed

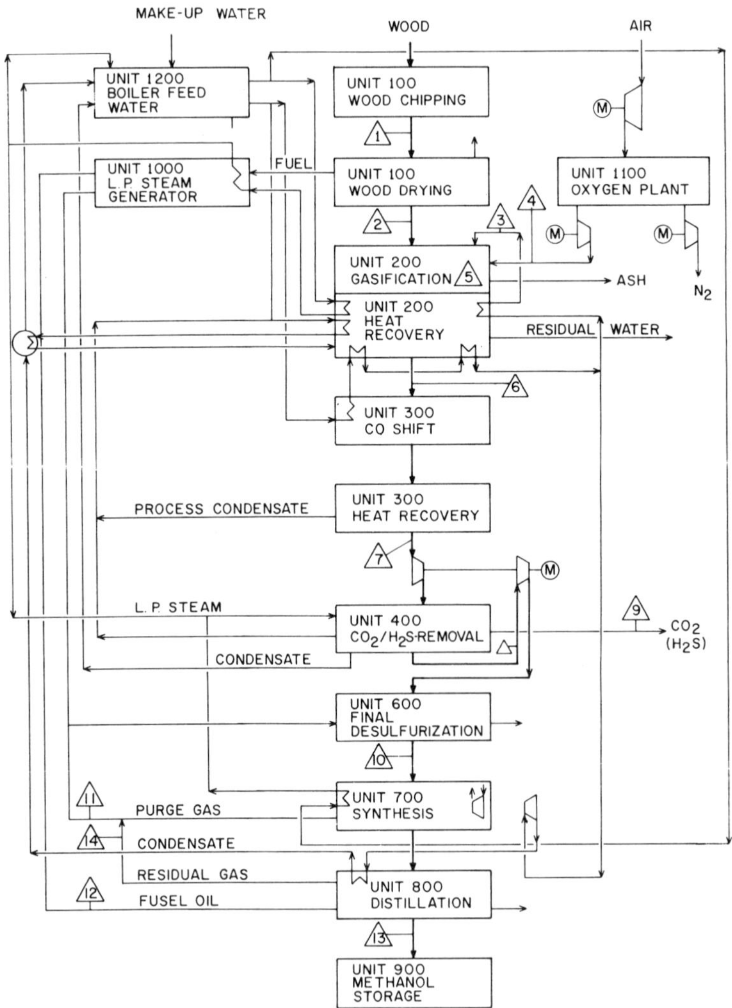

Figure 1.   Methanol synthesis—Winkler gasifier

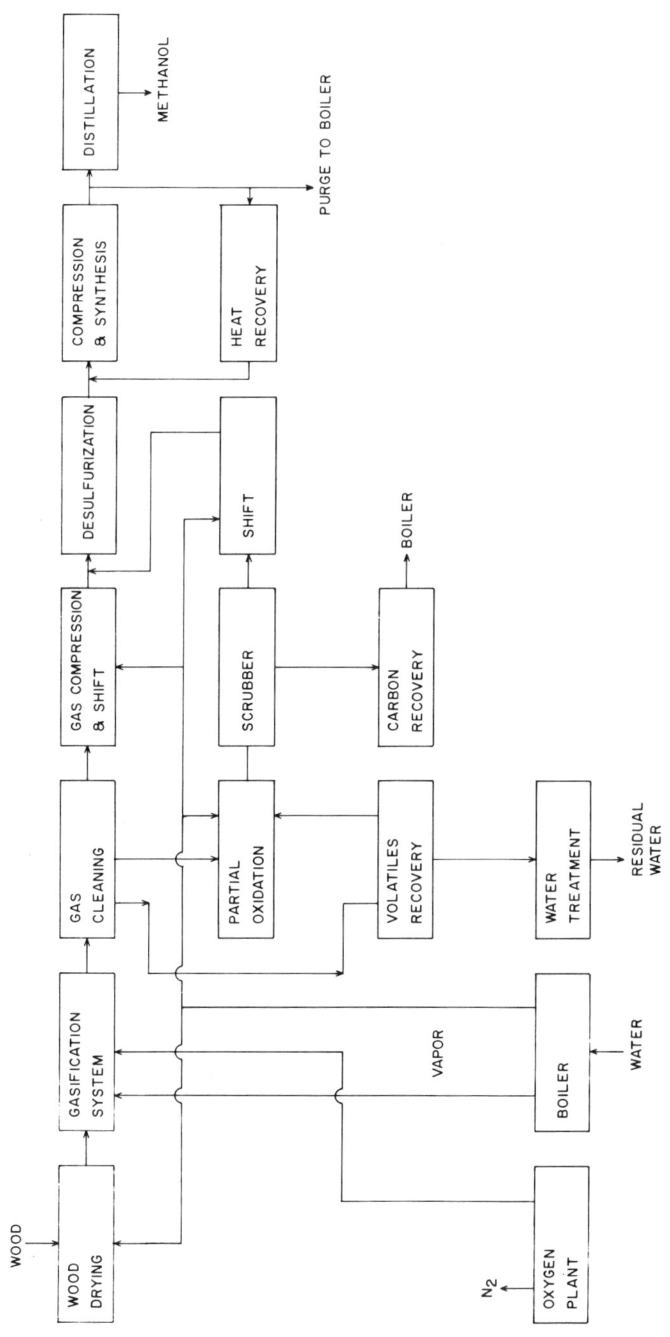

Figure 2.   Methanol synthesis—conventional gasifier

conventional gasifier that produces a gas containing large
amounts of tars that have to be washed out. The tars can be
used for steam generation or, depending on the thermal balance of
the plant, can be sent to a partial oxidation unit. All the
plant sections are conventional, except the gasification system.

Material and Energy Balance

A 1,000 ton/day methanol plant, using a Winkler type gasifier
adapted to burn wood chips and to use the ICI process for
synthesis, requires the following consumption of ingredients per
ton of methanol:
- wood (30% moisture) = 2.6 t
- electricity = 720 kWh
- cooling water = 115 m^3
- clean water = 1.5 m^3

If ones utilizes conventional gasifiers operating at
atmospheric pressure, such as the Davy or Lacotte type, raw
material consumption goes up. This also applies if charcoal is
used as a raw material. The main reasons are that much of the
tars is lost and more energy is needed to compress the gases.

	Conventional wood gasifier	Charcoal gasifier
Wood (30% moisture)	3.1 t	~ 4 t
Charcoal	-	1.1 t
Electricity	1,250 kWh	790 kWh
Cooling water	115 m^3	135 m^3
Clean water	1.5 m^3	1.6 m^3

Adopting 4,770 kcal/kg for the low heat value of methanol
and considering the case of a Winkler type gasifier, the energy
balance of the plant would be as shown in Table III:

Table III
Energy Balance of Methanol from Wood Plant

Item	Input Gcal/hour	Output Gcal/hour
Wood (30% moisture)	331.1	-
Water	0.7	-
Electricity	25.6	-
Compressed air	7.0	-
Methanol	-	198.8
Losses		
Residual air	-	23.7
Combustion gases	-	3.3
Cooling water system	-	45.0
Gas cooling system	-	75.5
Others	-	18.1
TOTAL	364.4	364.4

The net overall efficiency is around 55%, less than the figure for methanol plants using natural gas with efficiencies as high as 65%.

Fixed Investments and Raw Material Costs

<u>Investments</u>. Several studies have been published, evaluating the investment costs of methanol production from coal, lignite and biomass. The set of values have been selected as shown in Table IV:

Table   IV
Investment Costs of Methanol Plants

Reference	Raw Material	Capacity t/day	Total Investment Cost (US$ x 10^6)
(4)	wood	1,000	137
(8)	wood	500	64
(8)	wood	2,000	169
(7)	coal	2,000	330
(6)	coal	2,500	306
(6)	coal	1,000	150
(9)	coal	5,000	531
(9)	coal	1,000	141
(9)	lignite	5,000	630
(9)	lignite	1,000	178

From these numbers we may deduce that:
- Investment costs in methanol plants from wood, coal or lignite have the same order of magnitude. For a 1,000 t/day plant this cost is in the range US$ 140 to 180 million under present conditions in Brazil.
- Investment costs for methanol plants from wood ought to be less than for methanol plants from coal because wood has a very low ash content and is practically free of sulfur.
- Methanol production is not a "capital intensive" process as compared to other synfuel production systems. Fixed capital cost is between US$ 0.21 and 0.35 per liter/year of installed capacity. This investment cost is similar to that needed for ethanol production in Brazil.

The cost breakdown (4) is as shown in Table V:

Table V
Investment Cost Breakdown

Item	% of total cost
Process equipment	30.3
Auxiliary equipment	8.8
Off sites	2.2
Civil work	3.6
Erection	18.7
Engineering and administration	9.8

Licensing	1.5
Spare parts	3.0
Transportation and insurance	9.1
Interest during construction	13.0
	100.0

For a plant in Brazil it is important to verify that the
amount to be paid in foreign currency is about 30% of process
equipment and engineering, plus licensing costs. Total imports
will represent about 13% to 15% of the total investment costs.

   <u>Wood Costs and Productivity</u>. The F.O.B. costs for wood in
the States of São Paulo and Mato Grosso do Sul are as follows:

Costs (US$/m^3)

Item	São Paulo	Mato Grosso
Standing wood	4.00	1.96
Exploration	2.24	2.24
Transport	5.60	7.28
	(100 km)	(200 km)
TOTAL	13.60	11.48

   Wood transportation represents almost 50% of its F.O.B.
cost. Thus it is a very important item to be considered in site
selection. Assuming for wood a heat value around 3,500 kcal/kg
(30% moisture) the energy cost will be between US$ 4.68 and
5.55/Gcal or US$ 1.2 and 1.4 per million BTU.
   The wood costs above are very close to the price paid to
wood producers by the cellulose plants in the State of São Paulo.
   The wood cost per ton of methanol produced is between US$
42.8 and 64.2 (2.6 to 3.0 t of wood per ton of methanol). These
costs can be reduced if the plant is located inside the forest
itself.
   The average value of wood productivity (in m^3/ha per year)
is a function of the soil characteristics, seed genetics,
micro-climate, cycle of cutting, and other parameters. For the
average Brazilian situation, eucalyptus industrial forests,
planted in poor lands (cerrado, a type of savanna, for example)
and cut every 7 years, have a production of 20 t/ha per year
(50% moisture) which permits production of 5 to 7.7 tons of
methanol/ha per year.

   <u>Operating Costs</u>.
   a) Manpower: the required manpower can be defined as follows
for a plant with production capacity of 1,000 tons of methanol/
day:

- engineers	-	12
- technicians	-	193
- labor	-	295
TOTAL		500

The expenditure for the above mentioned group, including
social benefits, can be estimated at US$ 5.57 million per year.
The man-hour cost participation in the total production cost is
of the order of US$ 18.6 per ton.

b) Electricity: if the synthesis compressor and the oxygen
plant section are electrically driven, the total electrical power
requirement for the plant is 29.7 MWh/h for 1,000 t/day methanol
production. If no cheap electricity is available, it is
possible to use steam driven compressors. In this case wood
consumption is increased 15% and electric power is reduced to
7.1 MWh/h.

The above analysis assumes cost of electricity is US$ 30/MWh,
which is the average industrial electricity rate in Brazil.

c) Catalysts and chemicals: one may adopt US$ 1.5/t of
methanol, on the basis of international and Brazilian experiences.

d) Maintenance: for the type of plant considered, it is
reasonable to estimate maintenance costs as 2.5% of the total
investment, amounting to US$ 3.425 million per year.

Production Costs

From the preceding values, the following production costs
(US$/t of methanol) are shown in Table VI:

Table VI
Methanol Production Costs

Item	Wood Gasifier	Conventional Gasifier	Conventional Charcoal Gasifier
Raw material (wood or charcoal)	41.8	49.8	64.2
Electric power	21.6	37.5	23.7
Labor	18.6	18.6	18.6
Catalysts	1.5	1.5	1.5
Maintenance	11.4	11.4	11.4
TOTAL	94.9	118.8	119.4

Capital costs should be added to these costs, calculated
for an amortization period of 20 years, 10% yearly interest,
annual production of 300,000 t of methanol (plant service
factor = 82%), as indicated below:

Investment Level (US$ x 10^6)	Capital Costs (US$/t of methanol)
100	39
120	47
140	55
160	63
180	70
200	78

Consequently, total production costs (as a function of the

investment level and of the adopted technology) are shown in
Table VII:

Table VII
Total Production Costs of Methanol

Investment	Winkler	Gasification Process	
		Conventional (wood)	Conventional (charcoal)
US$ x 10^6	US$/t	US$/t	US$/t
100	132	158	158
120	142	166	166
140	150	174	174
160	158	182	182
180	165	190	190
200	173	198	198

As the required investment is between US$ 140 and 180 million
the cost of methanol from wood is between US$ 150 and 165/t using
advanced technology, and between US$ 174 and 190/t using
traditional gasification systems.

Methanol Costs and Oil Prices

There is no accurate way to determine the real costs of the
various oil derivatives. Oil refining is a continuous operation
and it is very difficult to allocate costs to each derivative
obtained. In Brazil this question involves certain social
criteria, aimed at reducing or increasing the consumption of a
given derivative, minimizing social effects of oil price
increases, fostering the development of some branches of industry,
etc. It is certain that the average price attributed to the
various derivatives should cover all refining costs, including
the capital costs of the refineries.

To compare costs of methanol and oil derivatives, it is
advisable to take the real costs of methanol and the average
refining costs of the various derivatives, while also considering
the corresponding heat contents and efficiencies in each use.
In an economical macroanalysis there is no sense in comparisons
based just on market prices.

In a very simple way, it can be stated that the production
cost of methanol should be kept below an upper limit (or maximum
acceptable cost) given by the expression:

$$Cm = \frac{Cp + Cr}{159} \times \frac{H.V.m.}{H.V.oil} \times \eta$$

Cm = maximum acceptable production cost of methanol
      (US$/liter)
Cp = F.O.B. price of oil (US$/Bbl)
Cr = refining and transportation cost of oil (US$/Bbl)
H.V. methanol/H.V. oil = heat value relationship
η = efficiency in the substitution of methanol for a given

oil derivative.

The following data might be considered:

Cp = varying from US$ 25 to US$ 50/Bbl

Cr = US$ 6/Bbl (average Brazilian costs)(10)

H.V. methanol = 4,700 kcal/kg

H.V. oil = 10,500 kcal/kg

η = depends on the derivative to be substituted.

- Partial substitution for gasoline: utilizing up to 15% methanol, compression ratios of the engines in Brazil could be increased. The fuel consumption does not follow the proportion based on the heat values. Practically the η value is around 1.4, similar to ethanol.

- Complete substitution for gasoline: in this case the gain in efficiency is not very high. The experience with ethanol indicates $\eta \simeq 1.2$.

- Diesel oil substitution: tests made by CESP/IMT (5) indicate that for producing 1 kWh net with diesel engines requires 11.5 MJ of diesel oil or around 10.5 MJ of methanol, which indicates $\eta \simeq 1.1$.

- Fuel oil substitution: $\eta \simeq 0.90$.

The values of Cm (methanol) are shown in Table VIII, in US$/t, for oil prices varying from US$ 25 to 50/Bbl, and various types of oil substitution:

Table VIII

Maximum Acceptable Methanol Production Costs

Oil price US$/t Substitution	25	30	35	40	45	50
Partial substitution	153	177	202	227	252	276
Complete substitution	131	152	173	195	216	237
Diesel oil substitution	120	139	158	178	198	218
Fuel oil substitution	98	114	130	146	162	178

At present oil prices, around US$ 30/Bbl, the Cm values are between US$ 114 and 177. As oil prices tend to rise and methanol costs remain practically fixed, methanol as a fuel becomes an economically feasible alternative under Brazilian conditions and its feasibility tends to improve as oil becomes more and more expensive and difficult to obtain.

CESP's Experiments in the Production of Methanol from Wood

CESP - Companhia Energética de São Paulo - has developed tests on charcoal and wood gasification, as well as methanol synthesis, in a pilot plant (Corumbataí) in Rio Claro, State of São Paulo. The plant produces 20 kg/hour of methanol using two charcoal gasifiers and has been designed to produce 1,000 kg/day of methanol directly from wood with a third gasifier that has been operating since September, 1980.

The main objectives of the plant are:
- to define operational, technical and reliability aspects
of charcoal and wood gasification
- to develop alternative technologies for wood and charcoal
gasification and synthesis gas treatment
- to promote alternative catalyst tests, defining their
main characteristics
- to evaluate Brazilian capability in equipment manufacture
and design, engineering and erection of gasification systems,
high pressure plants, methanol synthesis, utilities, etc
- to aid in training operators.

<u>Corumbataí Pilot Plant</u>. The flowsheet in Figure 3 shows the
main process diagram and its equipment.
A - Low pressure area
   . Gasifiers: two units using charcoal
                one unit using wood directly
   . Gas coolers: three units
   . Gas washers: one unit
   . Dewatering: one unit
B - High pressure area
   . Main compressor: three stages, 55 atm, 64 Nm^3/h capacity
   . Shift converter: made of stainless steel, operates at
     $500^{o}C$, iron oxide catalyst
   . Desulfurization unit: a stainless steel column, ZnO
     catalyst, $400^{o}C$
   . Methanol converter: two stainless steel columns, in
     series, operating at $300^{o}C$ and 50 atm, Cu catalyst
   . CO2 removal unit: made of carbon steel, operates with $5^{o}C$
     water in spray
   . Electric gas heater
C - Services
   . Electrical switchyard, 1,200 kVA
   . Electric boiler, for superheated steam
   . Water cooling unit (for the CO_2 removal unit)
   . Methanol storage tanks.

<u>Process description</u>. In both charcoal and wood gasifiers,
the carbon monoxide and hydrogen are obtained by passing steam
into a bed of charcoal (electrically heated). In the wood
gasifier, wood is converted into charcoal by recirculating the
hot gases and the wood volatiles from the top to the bottom of
the vessel. The hot gases carbonize the wood, and the tars are
destroyed by circulating in the hot area of the gasifier
(gasification area in the bottom which has a temperature around
$1,250^{o}C$) thereby generating more synthesis gas.
The final gas temperature is $85^{o}C$; it is further cooled to
$50^{o}C$ by direct contact with a water spray, and then washed; the
gas passes through a dewatering column and is compressed to 50
atm. Its composition at this stage is:

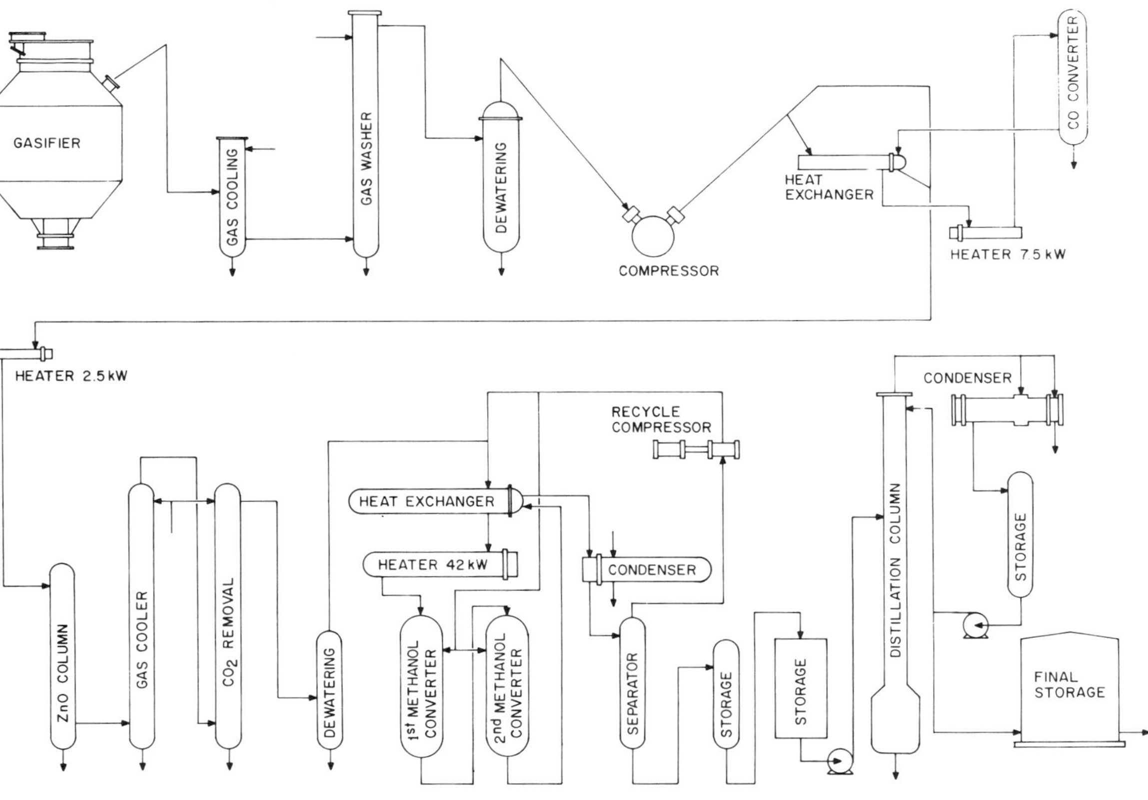

Figure 3.   Methanol synthesis—CESP's pilot plant

H_2 - 50%
CO   - 39%
CO_2 - 5%
N_2 - 5%
The synthesis gas composition, as required for the
installation, is:
H_2 - 67.4%
CO   - 18.8%
CO_2 - 9.4%
N_2 - 4.4%
To achieve this composition, part of the gas flows to the
shift converter and, reacting there with the superheated steam,
produces CO and CO_2. There is a heat exchanger to raise the
temperature of the compressed gas to the shift temperature; the
heat comes from the shift reaction. The gas is introduced in a
ZnO column, to eliminate sulfur traces, then is cooled with
water, and passes through the CO_2 removal column. The CO_2 is
absorbed by $5^{\circ}C$ water at 50 atm, and the gas is now ready for
synthesis.

Before entering the synthesis vessel, the gas is mixed with
the recycled gas which is then heated by exchange with synthesis
reaction heat; the gas enters the first reactor at $240^{\circ}C$ and
leaves the second at $270^{\circ}C$. In this exchanger the exit gas
exchanges heat with the gas that enters the synthesis reactor;
and afterwards it is cooled to $35^{\circ}C$, which condenses the methanol
vapors. The liquid methanol is separated and the non condensed
gas is recycled.

Just for start up, there are electric gas heaters before the
shift, the desulfurization and the synthesis units; during normal
operation, these heaters are not used.

Future developments. As mentioned earlier, the major
technical problem for the production of methanol from wood is
wood gasification. No wood gasifier for large scale gas
production is now available, because no one has really been
interested in this question for economic reasons. Technically,
wood gasification is easy: wood is a very homogeneous material,
has low ash and sulfur content, and behaves as a very good fuel.
There are various wood gasifiers that operate pretty well for
small amounts of wood. Historically, wood and charcoal have been
gasified for many decades, mainly in France and also in Brazil,
during the Second World War. What has to be done now is to
design, build and test some demonstration gasification plants on
a commercial scale, in order to define the main design data. Two
alternatives lines might be followed:
      - to design a completely new wood gasifier
      - to adapt some existing coal gasifier to operate with wood.
With a view of saving time, CESP decided on both alternatives.
After an in-depth survey of the available technologies,
three options have been made:

1. To develop and build a wood gasifier using experience from
the present pilot plant. This will be a fixed bed unit, with
recirculation of wood volatiles; the reaction heat will be
obtained partly from electric power and partly through oxygen
injection which permits good reaction control and avoids
excessive consumption of electricity.
2. To develop and build a wood gasifier and a gas cleaning system
using a very traditional technology. This is a very conservative
and prudent approach that certainly will give good results. Such
a gasifier (fixed bed, atmospheric pressure, counter-current)
will actually be the result of updating some units built in
India for an ammonia plant just after the Second World War.
3. To adapt for wood an advanced lignite gasifier that operates
co-current at 10 atm in a fluidized bed. This system has many
advantages, produces a very clean gas, but is now in a very
preliminary development stage.

Each of these envisaged units will produce an amount of gas
equivalent to 100 t of methanol per day, burning around 300 t of
wood per day. The engineering development for the three units
is just beginning and is being carried on by joint ventures of
Brazilian, American and German private companies, one for each
unit. All units are scheduled to operate by the end of 1981.

CESP's Experience in Using Methanol as a Fuel

Methanol can be used as an alternative fuel for gasoline,
diesel and fuel oil. Tests are being made for these three
alternative uses. The methanol fuel will substitute for oil
derivatives defined by Brazil's energy policy. What has to be
done now is to verify the technical advantages and disadvantages
to each form of utilization.

Gasoline substitution. Two General Motors vehicles are
operating in their normal work, using methanol as a fuel. One
has a 6 cylinder engine, the other a 4 cylinder engine.
Modifications were:
    - pre-heating of the air-methanol mix
    - more powerful ignitions
    - increase in the compression ratio (from 7.5:1 to 11:1)
    - substitution for the original carburetor
These vehicles have already run 35,000 km without any
problems at all. The 4 cylinder car fuel consumption is 7.3 km/
liter and the 6 cylinder (a G.M. pick-up) is 3.7 km/liter.
Other engines, mainly Volkswagen, are being tested in
dynamometers to measure the emissions (pollution control) and
other general performance data.
For gasoline substitution major problems are not expected.
Brazil has used ethanol blended with gasoline for years without
problems and now ethanol engines are part of the normal
production of the Brazilian car industry. Methanol is very

similar to ethanol as a fuel. However, by 1985 Brazil is
expected to produce ethanol equivalent to 60% of its gasoline
demand. Under these conditions, methanol will probably
substitute for diesel and fuel oil, the derivatives that are now
increasing Brazil's oil importation.

   <u>Fuel oil substitution</u>. A 15 MWe thermogenerator has been
remodeled to burn methanol. This type of boiler is currently
manufactured in Brazil and is widely used here. Its main
characteristics are:

<u>Type</u>	VU 60
Capacity of superheated steam	68,000 kg/hour
Steam temperature	440°C
Steam pressure	42.2 kgf/cm^2
Fuel oil consumption	6.4 t/hour
Number of burners	6 (A-20 type)
Efficiency	85%

The main modifications took place in the burners. The
boiler itself has not been modified. In the tests it achieved
60% of nominal output of steam. With better monitoring of
superheater temperatures, it is expected to achieve an output
of 90%. The test program will be developed this year (1980) and
will include emission analysis, boiler efficiency, corrosion,
and other aspects. CESP also plans to analyse smaller boilers
which will operate this same year.

   <u>Diesel substitution</u>. Diesel fuel is the basic fuel for the
Brazilian transportation system, and structural modifications
to this system are very difficult to implement. To substitute
for this fuel, several experiments are being carried on by CESP:
- use of additives: research on existing and new additives.
Minimum changes are required to adapt the engine. Several
vehicles and engines are being used for the tests.
- dual fuel engines: a diesel-electric locomotive operates with
methanol and diesel oil. This process is suitable for heavy
vehicles requiring two fuel tanks. Consumption of diesel oil is
reduced by 50% in preliminary tests at nominal output.
- high compression ratio engines: a four cylinder Detroit Allison
engine with a compression ratio of 24:1 was installed in a
truck and is operating in normal service, using 85% methanol and
15% castor oil. This engine has operated eight months without
any serious problem.
- hot spot: very good results are being achieved by inserting a
hot spot (electrical resistance) in the combustion chamber of
the diesel engine. Methanol with only 1% castor oil (for
injection pump lubrication purpose) has shown good performance,
even on cold start.
   All of these experiments are in preliminary stages of
development. The results obtained until now are very good, but
detailed information on fuel consumption, life of the engine,
economics, average emission rates, etc will be available only
after completion of the experiments mainly at the end of 1980.

Literature Cited

1. "Balanço Energético Nacional", Ministério das Minas e Energia, Brasília, 1978.

2. "Modelo Energético Brasileiro", Ministério das Minas e Energia, Brasília, 1978, p. 12.

3. Silva, J.G.; Serra, G.E.; Moreira, J.R. and Gonçalves, J.C. "Balanço Energético e Cultural da Produção de Álcoois de Cana-de-Açúcar, Mandioca, Pinus e Eucaliptus", IFUSP, São Paulo, 1979.

4. "Metanol: Estudo de Viabilidade", Inter-Uhde Engenharia Química, São Paulo, 1978.

5. Nanni, N.; Domschke, A.G.; Brunetti, F.; Rozov, V.; Rosa, F.; Borba, M.L.O. and Abreu, R.S. "B-39 Use of Glow-Plugs in order to Obtain Multifuel Capability of Diesel Engines", Instituto Mauá de Tecnologia, IV International Symposium on Alcohol Fuels Technology, Guarujá, 1980.

6. "Gasification of Brazilian Coal Using the Koppers-Totzek Process - Methanol Plant", Krupp-Koppers, Rio de Janeiro, 1976.

7. "Informações Técnicas e Econômicas Preliminares da Utilização de Gaseificação de Carvão para Produção de Amônia, Metanol e Gás para Redução Direta", FINEP, Rio de Janeiro, 1976.

8. Katzen, R. "Chemicals from Wood Wastes", U.S. Department of Agriculture, 1975.

9. "On the Trail of New Fuels - Alternative Fuels for Motor Vehicles", Ministry of Research and Technology, West Germany, 1975.

10. Bellotti, Paulo "A Execução da Política dos Combustíveis Líquidos e Gasosos", I Congresso Mackenzie de Minas e Energia, São Paulo, 1978.

RECEIVED February 17, 1981.

Ethanol: Manufacture and Applications

I. B. MARGILOFF, A. J. REID, and T. J. O'SULLIVAN

Publicker Industries Inc., 777 West Putnam Avenue, Greenwich, CT 06830

The production of ethanol via fermentation is in many cases easier done than avoided. Few if any primitive tribes are without their alcohol beverages based on local sugars or starches, and residues from beer have been discovered in the Pyramids. Thus it is impossible to set a time when production of ethanol for beverage purposes began. Most alcohol in undeveloped countries is still for beverages. Aside from beverage production, ethanol (ethyl alcohol or "alcohol") is made for industrial purposes by both synthetic and fermentation processes.

Synthetic Alcohol Manufacture

The first synthetic ethanol was produced by a subsidiary of Union Carbide at South Charleston, West Virginia in the late 1920's; thus synthetic production is over 50 years old.

In the original process, ethylene gas was fed into strong sulfuric acid (around 90 per cent) at moderate temperatures and near atmospheric pressure. The ethylene was converted largely to ethyl hydrogen sulfate, but partly to diethyl sulfate. Some polymerization of the ethylene also took place.

The rich acid was then diluted with water and the temperature raised, hydrolyzing the esters and driving off the alcohol thus produced, along with ethyl ether and the lower boiling polymers.

The dilute acid was then concentrated at atmospheric pressure, producing a reusable acid but also a number of troublesome by-products. Sulfur trioxide vapor was produced as the acid became more concentrated. This produced an intolerable fog, so electrostatic precipitators were installed and the remaining vapor was decomposed by heat into sulfur dioxide which was discharged from a high stack.

Much of the polymerized ethylene could be removed from the weak acid by skimming and filtration, but a certain amount remained. This was partially oxidized by the acid during concentration, producing carbon dioxide and carbon; sulfur dioxide was another product of the concentration process.

Vacuum concentration of the acid using Mantius, Swenson and other concentrators was used to some extent, but presented operational difficulties.

Although a patent was issued as early as 1936 covering aqueous phosphoric acid[1] the sulfuric acid ester process was unchallenged for nearly 20 years in spite of corrosion and pollution problems, but is no longer in much use.

Current synthetic production uses a weaker non-oxidizing acid. A respectable reaction rate requires higher temperatures, which shifts the equilibrium unfavorably, and high pressures. Typically, a reactor is operated using phosphoric acid on a porous carrier with a temperature around 250° C and a pressure of 70 bars. The Shell process uses granular silica gel, while Veba-Chemie (now part of Chemische Werke Huels) uses balls apparently made of clay which is leached after firing. The silica gel apparently produces more conversion per pass, while the Veba catalyst is far more durable. "Case hardening" of silica gel[2] greatly improves durability, and the Davison division of W. R. Grace & Co. has an experimental pelleted catalyst carrier based on silica gel which is claimed to be even better.

Typically, the hydration occurs using 0.3 - 0.5 moles water per mole ethylene and a space velocity around 1 min $^{-1}$, with 4 - 5 per cent conversion of ethylene per pass. The reaction is exothermic but because of low conversion the reacting mixture does not require cooling. Ether is produced but may be recycled if it is not wanted as a by-product.

Increased ethylene concentration or temperature produces more oils, butanes, carbon, etc., while more water dilutes the phosphoric acid and reduces its activity. Exact operating conditions will vary with the condition of the catalyst.

There is some loss of phosphoric acid into the product leaving the reactor in the form of an ethyl ester or perhaps as some species of phosphoric acid. It is replaced by either continuously or intermittently adding makeup acid or by shutdowns and reimpregnation.

It is not impossible that other acids could be used instead of phosphoric. One would look for an extremely non-volatile acid, probably a transition metal oxide with good resistance to reduction and low water solubility (to permit more water in the feed). One producer may be using such an acid.

Following reaction, the ethylene gas is scrubbed to remove the alcohol and more or less of the ether and is then recycled. A side stream is purified, vented, or used in another process.

A crude synthetic alcohol is obtained at 15 - 30 per cent alcohol and contains acetaldehyde, oils, butanols, ether and some dissolved C_4 - C_6 hydrocarbons along with ethylene. It is distilled to remove low and high-boiling impurities and to concentrate to 190 proof. Details of the distillation are often proprietary. A good "spirits" grade of synthetic alcohol is at least as pure as the best fermentation alcohol.

Small amounts of synthetic alcohol occur as by-products in various chemical processes, particularly production of acetic acid via butane oxidation. One company is producing about 12 million gallons of by-product ethanol at present but is not recovering it.[3]

Fermentation Alcohol Manufacture

United States capacity for fermentation industrial ethanol is estimated at 80 - 90 million gallons per year but this could be increased by diversion of distilled beverage production. About 10 million gallons of the fermentation alcohol is derived from wastes such as sulfite liquor or whey, but the balance presently depends on damaged grain or sugar, crop surpluses and molasses. Relative costs of ethylene and carbohydrates will change this in a few years, but no reversal to ethanol-based ethylene is likely in this country for much longer, perhaps through 1990.[4]

There is curently a government-sponsored program to increase fermentation alcohol capacity to provide material for gasohol. The issues here are numerous and complex, and emotionally-colored evaluations are clearly evident even among those who profess to have purely scientific or technical interests. Barring a drastic rise in grain prices the program is likely to continue.

The basic process for fermentation alcohol is little changed over the last hundred years. Typically the grain is ground, slurried with water and cooked. The cooking process causes the starch granules to take in water, swell and eventually rupture. This releases the starch molecules into the mash as long chain dextrins which are in turn broken down by the action of a saccharifying enzyme to simple sugars. Yeast is added to the mixture, and the sugars are converted to alcohol and carbon dioxide. The fermented mash, now called beer, is stripped of its alcohol by distillation. The spent beer which consists of the non-fermentable portions of the grain is separated into two fractions, the solubles and insolubles. The soluble portion is evaporated to produce a syrup containing 35 - 50 per cent total solids. This can be marketed as is or dried with the insoluble portion to form distillers dried grains with solubles---DDGS. The insoluble fraction can be marketed wet or dried to form distillers dried grains---DDG. These by-products are usually sold as dairy ration supplements, especially the wet grains and whole stillage fractions.

Molasses fermentation is more straightforward, as the milling and saccharifying operations are not needed. The molasses is diluted to 15 - 20 per cent sugars, depending on conditions, fermented, distilled and the remainder evaporated to a syrup. However, this syrup can only be used in small proportions for

feeding because of its high potassium content. Alfa-Laval[5] is
promoting a process for reducing this potash level in the syrup
and obtaining potassium for fertilizer use and return to the
land.

The sulfite process for paper-making produces a certain
amount of wood sugars which cannot be sewered because of pollution laws. Fermentation permits recovery of the sugar as alcohol
and single cell protein. A commercial plant is in operation.

An economically ideal substrate for fermentation is cellulose. However, it is associated with lignin which inhibits
hydrolysis of the cellulose to simple sugars. The earliest
studies of ethanol production via hydrolysis of cellulose were
carried out in Germany during World War I. At that time two
proceses were developed; the weak acid Scholler process and the
strong Rheinau technique. Studies on the enzymatic hydrolysis
of cellulose were initiated during World War II to try to find
a remedy for the microbiological degradation of soldiers' uni-
forms. However, it is only within the last ten years that a
renewed interest in hydrolysis of cellulose has taken place and
cellulase enzymes are now being developed.

Variations of acid hydrolysis processes are being investigated. Certain fungi, especially _Trichoderma reesei_, can also
degrade cellulose to sugars. To date, most results of cellulose-
type enzymatic degradations produce only low sugar concentrations
(e.g. 50 - 60 per cent yield at 5 - 10 per cent glucose[6]).

The technology---though not the basic process---associated
with fermentation has developed considerably over the last fifty
years. One of the major breakthroughs was the Melle process for
the reuse of yeast. This was developed by Les Usines de Melle[7]
and incorporated the use of centrifuges to recover yeast from
the beer prior to distillation. This yeast was in turn used to
inoculate the next fermentor. The process has three major
advantages:
   1. the efficiency of fermentation increases, reaching the
       Pasteur theoretical,
   2. fermentation takes place at a faster rate, and
   3. the cost of yeast replacement is considerably reduced.
The acclimation of the yeast cells to the particluar substrate
combined with the large numbers account for the first two advantages, and the third can be costed out on the basis of yeast cost
versus purchase and operation of centrifuges.

Cell recovery is mainly concerned with what is known as
clear mash, i.e. sugar solutions without any entrained solids
suchas as those in a whole-grain mash. The use of cell recovery
on mash can be accomplished by clarifying the mash prior to
fermentation or else incorporating a wet-milling front end and
taking the saccharified starch as substrate. However the production of a clear mash directly from cereal mash is presently not
feasible. This is due to excessive losses of carbohydrate in
the cake or non-fermentable extract ranging from 10 to 20 per

cent of total starch. This cake even, with a higher carbohydrate
content than DDG, is not as valuable as DDG because of its lower
protein concentration.

Taking the idea of yeast recovery one step further is continuous fermentation. This incorporates a vessel or vessels
inoculated with yeast through which the fermentable substrate
is passed continuously. The flow is designed to ensure that all
the sugar is fermented before it exits the final fermenter as
beer. This beer is clarified to remove any entrained yeast cells
and a percentage of the yeast is recycled back into the system.
The excess can be further processed depending on the market
application, i.e. for human or animal use.

A great amount of time, money and effort is being devoted to
the use of cellulose as a feedstock for the production of ethanol.
The studies incorporate chemical or enzymatic conversion of the
cellulose to glucose and the conversion of this to ethanol with
yeast (Saccharomyces) or bacteria (Zymomonas). However, a third
process is presently under development at Massachusetts Institute
of Technology whereby the direct conversion of cellulose to ethanol is being attempted without a separate hydrolysis step.[8]
This is accomplished by utilizing two separate microorganisms
which work symbiotically with each other, Clostridium thermocellum and Clostridium thermosaccharolyticum to produce ethanol.
The research is still in its infancy and several years of research and development are anticipated before its use commercially.

Within the last ten years phenomenal advances in the field
of genetic engineering have taken place. It is hoped, but not
yet demonstrated, that gene splicing may lead to major advances
in the production of alcohol, but carefully controlled use of
conventional microorganisms is now the best potential route to
good economics.

Applications

Synthetic alcohol demand in the United States in 1979 was
estimated at 208 million gallons (190 proof).[9] The same source
lists 94 million gallons for chemical manufacture, 42 million
for toiletries and cosmetics, 25 million for detergents, flavors
and disinfectants, 23 million for coatings and 25 million for
other uses. Synthetic alcohol is not used in beverages or gasohol (motor fuel).

Fermentation alcohol has been used almost entirely for
beverage purposes in less developed countries. In industrialized
nations, even with available ethylene, fermentation alcohol is
produced from low-cost surplus and waste materials. With the
shift in economics it will compete with synthetic in the latter's
fields. It has a monopoly of "gasohol" for political reasons
and of beverages for emotional reasons. Fermentation alcohol
capacity in the United States is uncertain but is estimated at

about 100 million gallons per year apart from beverages (65
million gallons produced in 1978---capacity not available).

As many as 70 products were at one time produced commercially from ethanol. Some of these downstream products are
butanol, 2-ethyl hexanol, crotonaldehyde, butyraldehyde, acetaldehyde, acetic acid, butadiene, sorbic acid, 2-ethylbutanol,
ethyl ether, many esters, ethanol-glycol ethers, acetic anhydride, vinyl acetate, ethyl vinyl ether, even ethylene gas.
Many of these products are now more economically made from other
feedstocks such as ethylene for acetaldehyde and methanol-carbon
monoxide for acetic acid. Time will tell when a revival of
biologically-oriented processes will offer lower-cost routes to
at least the simpler products.

Recent Developments

The outstanding recent development in ethyl alcohol is
"gasohol", a blend of fermentation ethanol (10 per cent) with
gasoline. It was studied thoroughly about 50 years ago[10,11] but
forced out by cheap oil. The alcohol must be nearly 200 proof,
especially in cold areas. Apart from extending the petroleum
supply directly, it permits use of lower-base octane gasoline
and thus indirectly further increases the yield of gasoline per
barrel of crude oil. Cooler engine temperatures reduce engine
wear and pollution. Views on gasohol are held, often quite
emotionally, and split into two camps. The opponents claim
the energy required to produce the alcohol is greater than the
energy it contains. They ignore the fact that gasohol seems to
give the same gas mileage as gasoline in spite of a lower Btu
content per gallon and that gasoline itself is subject to the
same criticism. Their energy values are suspect because existing distilleries' poor energy economics are a legacy of the age
when energy was cheap; there has been no demand for better technology until recently. The proponents of gasohol are often
oriented toward the agricultural benefits of increasing the
demand for cereal grains and other fermentables. The middle
ground is clearly to be preferred; some alcohol will be made for
motor fuel out of economically available substrates, but we
cannot realistically expect or demand that _all_ motor fuel sold
in this country have ten per cent fermentation alcohol based
solely on foodstuffs.

Several suppliers of motor fuel are considering contracts
and joint ventures with alcohol producers or building plants
independently. One view would suspect the choice of positions
may reflect a company's ability to produce premium unleaded
gasoline in desired amounts or its crude-oil position.

A second recent development is increased interest in the
conversion of cellulose to fermentable sugar for conversion
to alcohol. The enzymatic conversion is thus far not well
enough developed to be successful commercially, although break-

throughs may come any day. This new interest has been spurred
on by diminishing petroleum stocks and rising prices. It is
envisaged that cellulose, the most abundant polysaccharide in
nature, will fulfill part of our liquid fuel needs. Acid hydrolysis is feasible, in fact was practiced in Europe during World
War II to provide motor fuel, but this route is now no more than
marginally profitable. The readjustments of energy and raw
material costs now taking place will inevitably bring about
desirable technical developmetns including more favorable energy
balances. Use of straight 190-proof alcohol (95 volume per
cent) as an automobile fuel is being pushed in Brazil and may
find a place in less temperate climates as well. The need for
production of anhydrous alcohol for gasohol has stimulated a
search for new methods to replace the conventional ternary azeo-
trope method based on benzene or cyclohexane. Adsorption and
extractive distillation along with other azeotroping agents and
operation at various pressures are also being mentioned.

A few other comments: Finally, it should be noted that commercial production of
ethanol by continuous fermentation of grain mash has reportedly
been achieved at Archer-Daniels-Midland Company and that National
Distillers' researchers have also developed such a process.
Although continuous fermentation, per se, is not new there is
some skepticism concerning the degree to which this represents
a real technical and economic advance.[12]

A few other comments: Ethanol is listed by OSHA as a suspect but unproved carcinogen, but OSHA does not have juris-
diction over beverage alcohol, nor, since it is a natural product, does the Delaney Amendment apply. This could lead to
paradoxical situations. Ethanol will continue to grow explosive-
ly in volume in the next few years as a motor fuel; indeed,
this has happened in Brazil. The field is extremely volatile,
and new developments occur every week, particularly with respect
to fermentation and gasohol. Things should be clearer in about
five years, and we should remind ourselves to take a similar
look at ethanol then.

Literature Cited

1. U.S. Patent 2,050,445; August 11, 1936
       "Manufacture of Ethyl Alcohol", Floyd J. Metzger
          Assignee: Air Reduction Company, New York, New York

2. U.S. Patent 3,914,721; October 21, 1975
       "Combination Specular-Diffuse Projection Device and
       Method", John S. Pollock, Rochester, New York
          Assignee: Eastman Kodak Company, Rochester, New York

3. *Chemical Marketing Reporter*, July 16, 1979, Vol. 217, No. 3
    Chem Profile, page 9.

4. Anon. *Chem Week*, 1980, Vol 127, No. 19, page 41

5. Company literature, Alfa-Laval

6. "Molasses & Industrial Alcohol", Development Centre, OECD,
    Paris, 1978

7. Canadian Patent No. 341,720; May 15, 1934
       "Alcoholic Fermentation", Les Usines de Melle
    Canadian Patent No. 402,847; February 10, 1942
       "Alcohol Production by Fermentation", Les Usines de Melle
    Canadian Patent No. 348,549

8. Wang, D.I.C., R. J. Fleischaker and G. Y. Wang, *Chemical
    Engineering Progress Symposium Series*, 1978

9. Johnston, P.J., Ethanol: *An Alternative to Its Use as Fuel*,
    Workshop on Fermentation Alcohol in Developing Countries,
    March 26 - 30, 1979

10. Christensen, L.M., Hixon, R.M., Fulmer, E.J., "Power Alcohol
     and Farm Relief", *The Deserted Village*, No. 3; Iowa State
     College, Ames, Iowa, 1934

11. "Use of Alcohol from Farm Products in Motor Fuel", Letter
     from the Secretary of Agriculture to the Senate, May 3, 1933,
     Senate Document No. 57, 73rd Congress, 1st Session

12. Bishop, J.E., *Wall Street Journal*, November 14, 1980, p. 27

RECEIVED March 3, 1981.

Ethanol in Motor Gasoline

TED TARR[1]
Office of Alcohol Fuels, U.S. Department of Energy, Washington, DC 20585

J. R. JONES
TRW Energy Engineering Division, McLean, VA 22102

Transportation fuels derived by blending of biomass derived
ethanol and gasoline offer immediate potential to significantly
supplement petroleum derived fuels. Ethanol can supplement the
supply of gasoline produced from foreign crude oil. Oil imports
currently provide the raw material for production of half of the
liquid fuels consumed in the U.S., and represent a cash outflow
of almost $9 million per hour. Events in recent years have
dramatically illustrated the substantial economic cost, supply
vulnerability, and resulting economic instability resulting from
the high degree of dependency on imported oil. Biomass derived
ethanol offers an immediate method of reducing oil imports since
it is one of a very limited number of alternative fuels that is
likely to be available in significant quantity before 1985.

Alcohols fuels have been under detailed study in the United
States, Brazil, West Germany and other countries. In the United
States, evaluation of the use of ethanol as an automobile fuel
dates back to the earliest years of automotive use. For example,
the U.S. Department of Agriculture published a report in 1907
entitled "Use of Alcohol and Gasoline in Farm Engines"(1).
Interest in ethanol from grain has continued to the present day.
As an example, by late 1974 the Nebraska Agricultural Products
Industrial Utilization Committee had started sponsorship of a
gasoline-ethanol blend fleet testing program. Since the initial
testing, a broad-based grass roots movement has developed to
support the development of ethanol as a renewable alternative
domestic liquid fuel.

Much recent attention and investigation has been focused on
the use of ethanol as a gasoline extender, octane enhancer, or as
an alternative fuel. Many studies have been performed to evaluate the engine performance, emission characteristics, and the
advantages and problem areas encountered in conventional spark

[1]Current Address: Vulcan Cincinnati Inc., P.O. Box 86,
Mt. Airy, Maryland 21770

ignition engines when ethanol is used as a fuel. These studies
have defined the performance of ethanol and gasoline in various
proportion blends in conventional engines. Fleet testing of
various blends is currently being performed. Gasohol, which is a
mixture of 10 percent by volume anhydrous ethanol in 90 percent
unleaded gasoline, has expanded to include more than 2000 outlets
in 35 states.

On 11 January, 1980, the Administration announced an
expanded alcohol fuels program which will provide substantial
stimulus to accelerate domestic production of alcohol fuels from
sources other than petroleum. A target was set for domestic
production capability of 500 million gallons during 1981. If
this amount of ethanol were all blended to make gasohol, gasohol
would then account for almost 10 percent of our anticipated 1981
demand for unleaded gasoline or about one percent of total gasoline use. This represents approximately a six-fold increase
over existing alcohol fuel production capacity. The program
includes a variety of incentives, from tax credits to loan and
loan gurantees, which focus on two main objectives:
  o First, to permit gasohol to become economically
      competitive with unleaded gasoline at the pump
  o Second, to stimulate new investment in facilities to
      produce ethanol
These initiatives are designed to greatly increase the use
of gasohol in passenger and commercial vehicles, as well as to
encourage increased use of alcohol as fuel for off-highway use,
such as in farm equipment. Key elements of the expanded program
include:

*Permanent exemption for gasohol from the 4¢/gallon federal
gasoline excise tax.* The President originally proposed an exemption of gasohol from the 4¢/gallon federal excise tax in 1977.
In November 1978, the current exemption which expires in 1984 was
signed into law. In April 1979, the Administration proposed that
the exemption be made permanent in order to provide a long-term
incentive which investors in ethanol plants can count on over the
life of their facilities.

The Senate version of the Windfall Profits Tax bill would
extend the exemption to the year 2000. This exemption provides a
subsidy equal to 40¢ per gallon of ethanol. In concert with the
tax credit discussed below, it is the most important incentive
available to accelerate alcohol production and use.

40¢/gallon production tax credit. The Administration
supports the establishment of a tax credit for producers who use
the alcohol directly without blending with gasoline. This use is
expected to occur mainly on farms. The Senate version of the
Windfall Profits Tax bill would provide a production tax credit
of 40¢/gallon for alcohol over 190 proof, 30¢/gallon for alcohol
from 150 to 190 proof.

$3 billion federal credit program. A $3 billion, ten-year program of loans and loan guarantees for construction of small and medium-scale alcohol and other biomass production facilities has also been proposed. The smaller facilities would be located mainly on individual farms for use in farm equipment. This program will also assist farmer cooperatives for the production of alcohol for either on-farm or commercial use. In general, plants eligible for this assistance will produce less than 30 million gallons per year, with most facilities located on farms producing less than 5 million gallons per year. Funding for this program would be authorized at an annual level of $300 million, $250 million for loan guarantees and $50 million for loans, and will be administered through the Departments of Agriculture and Energy.

Energy Security Corporation programs for biomass. The Administration supports the allocation of up to $1 billion of assistance available through the proposed Energy Security Corporation for the construction of plants for the production of ethanol from biomass. The ESC will have a number of financing tools, including loans, loan guarantees, price guarantees and purchase agreements to encourage private investment in facilities with significant potential to reduce imports. The Senate bill, S. 932, includes such a provision and prompt Conference Committee action on this legislation is expected.

Revision of the entitlements program. The Department of Energy has revised the crude oil entitlements program to include ethanol produced from biomass. This provides an incentive currently equal to about 5¢ per gallon of ethanol used in gasohol. Since the entitlement program phases out along with crude oil price controls -- ending on September 30, 1981 -- this program offers its benefits to those who begin production soon, thereby accelerating our near-term use of gasohol.

10% investment tax credit. The Energy Tax Act of 1978 authorizes a 10% additional investment tax credit for equipment to produce liquids or gases from biomass sources including the production of alcohol. This tax credit is in addition to the existing 10% investment tax credit, for which alcohol production facilities are also eligible. Under current law this credit will expire in 1982.

Alcohol production research and development. The Department of Agriculture and the Department of Energy administer research and development programs to improve our methods for producing alcohol fuels from biomass and to broaden the range of biomass products which can be used to make alcohol. In FY 1980, a total of over $30 million is dedicated to this effort. This R&D is important for improving the competitive viability of alcohol fuels as well as the net energy balance.

500 million gallon target. The President announced a
national target for alcohol production capacity -- 500 million
gallons annually during 1981. Substantial amounts of alcohol
will probably be burned directly, particularly for on-farm uses,
rather than used for gasohol for commercial sale.
   Today, without incentives, production of ethanol is not
economically attractive when compared to gasoline, although
rising world oil prices continue to improve the economics of
gasohol. Net production costs for ethanol (after credits for
sale of by-products) are approximately $1.30 per gallon compared
to wholesale unleaded gasoline prices of $0.85 - $0.95 per
gallon. (Jan. 1980 values). With the various federal subsidies
in place, however, it is expected that an equivalent federal
subsidy of almost $0.50 per gallon of alcohol will be available
to producers. With this program combined with subsidies already
available in over half of the states, the resulting economics for
ethanol production are even further improved.

Ethanol Fuels Policy Issues

   Existing and proposed federal and state incentives for
fermentation ethanol production and use have contributed to the
rapid expansion of the gasohol market. In addition, a broad
spectrum of options is currently being pursued at the federal
level to help accelerate the commercialization of gasohol by
stimulating both its production and uses. Maximizing ethanol
production will require a mix of various sized ethanol plants.
Because of the lead time involved in building and operating
larger facilities, the Department of Energy and the Solar Energy
Research Institute (SERI) have produced a guide to provide basic
information to individuals interested in constructing small-scale
facilities (2).
   Several policy issues must be addressed when considering
production of significant quantities of ethanol for automotive
fuel use; these include:
   o The potential near-term demand for biomass-derived
      ethanol
   o The availability of raw material feed for ethanol
      production, and the degree to which ethanol fuel
      production would infringe on food production
   o The overall ethanol production economics, including
      sensitivity to such factors as the cost of raw feed
      material, scale of production, and co-product credits
   o the effect of state and federal incentives on the
      production economics
   o The net energy gain achieved when producing ethanol
   o The environmental impacts of ethanol fuel production and
      the automotive emission characteristics of ethanol-
      gasoline blends

Potential Demand for Biomass Derived Ethanol, 1980-1995.
Review of annual gasoline demand for use in passenger cars and
projected demands from various sources (3,4) indicate continuing
strong demand for gasoline through the 1980-1995 period as shown
in Figure 1. The reduced automotive consumption of gasoline,
resulting primarily from Corporate Average Fuel Economy (CAFE)
standards imposed on automobile manufacturers, will reduce annual
gasoline demand during the 1980-1990 period. After 1990, the
growth in the total number of automobiles will again become the
predominant effect and overall passenger car gasoline demand will
rise. In the near-term the level of imported oil will remain
high, and ethanol blended with gasoline will continue to offer a
method of reducing substantial oil imports.

Availability of Feedstocks. In the long term, if ethanol
fuel conversion capacity exceeds levels readily sustained from
surplus and distressed grains, and if cellulose-to-ethanol or
coal-to-methanol technology should be slow to develop and be
built, then it may be worth considering the encouragement of
additional crops (including energy crops, such as sweet sorghum,
which need not compete with food, feed, and fiber) or the use of
set-aside acreage for alcohol crop production. It appears that
an upper limit of approximately 3.3 billion gallons per year
(216,000 barrels per day) of ethanol could be produced from raw
material supplies using existing technologies, if conversion
capacity capable of processing these feedstocks existed (5). The
contribution of various agricultural commodities is shown in
Table I. This limit could be achieved by bringing into production all existing grain land and by supplementing food processing
by-products with sugar surpluses. However, achieving this limit
would be expensive, and would reduce the flexibility of U.S.
agricultural land and restrict options for food production. This
table shows the quantity of ethanol which could be produced if:
(a) USDA eliminates all future set-aside and diversion programs,
and all existing grain land is brought into productive use;
(b) cane sugar surpluses are converted to ethanol, (c) no new or
marginal cropland is assumed to be brought into production, and
sweet sorghum potential is not included.

Grain Alcohol Fuels Process Economics. As of late 1978,
the posted price by refiners and operators for unleaded regular
gasoline was in the range of $0.45-0.50 per gallon. During the
same period, the production cost of ethanol from corn was estimated to be from $1.05-1.16 ($1978). This would yield a dis-
counted cash flow rate of return of 15-20 percent with corn feed
cost of $2.30 per bushel and a by-product credit of $110 per ton
of distiller's dried grains. From December 1978 to December
1979, the price of gasoline had increased dramatically from
$0.45-0.50 per gallon to the range of $0.70-0.85 per gallon.

While the cost of producing ethanol has undoubtedly
increased during the past year due to increased capital and
operating expense, the price differential between unleaded

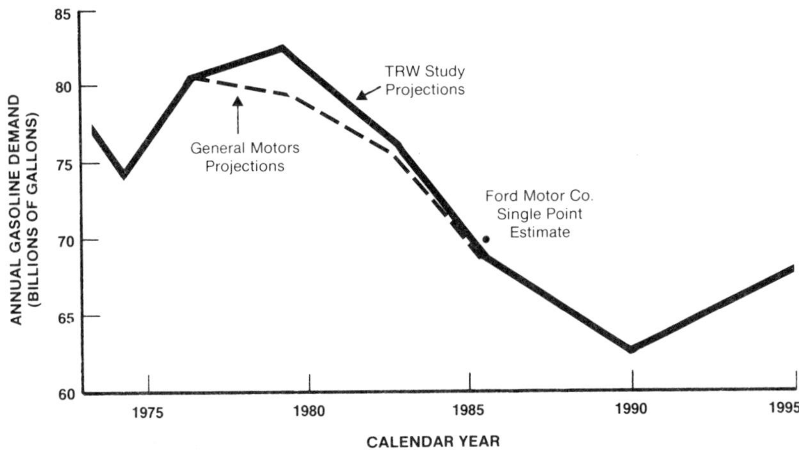

*Figure 1.   Comparison of annual gasoline demand projections for passengers cars
to 1995*

regular gasoline and grain ethanol motor fuel has narrowed
dramatically in recent months. Grain ethanol fuel is becoming
increasingly cost competitive as the price of gasoline continues
to rise at a rate greater than general inflation.

Existing analyses indicate that the production cost of grain
ethanol has a high degree of sensitivity to the cost of grain
feedstock and the credits claimed for by-products such as the
distiller's dried grain (6,7). As a rough guide, the cost per
gallon of ethanol must rise 12 cents for each 10 percent increase
in corn purchase price. Further, the cost per gallon could be
reduced 4 cents for an increase of 10 percent in the by-product
credit for distiller's dried grain. These approximate numbers
assume a new 50 million gallon per year ethanol plant with a
20-year life requiring an investment of $58 million (1978) at 20
percent return on investment.

Federal and State Subsidies. The large increase in demand
for gasohol is believed to be based, at least to some degree, on
consumer preference for vehicle fuel derived from renewable
resources, and the perception of the fuel as a high quality motor
fuel. In addition, increased demand has occurred because grain
ethanol for use in gasohol fuel mixtures has been made cost competitive with unleaded gasoline through state and federal price
incentives. Alcohol fuels currently receive a wide range of
federal and state government incentives. These incentives substantially improve the cost competitive position of grain ethanol
when used in gasohol, relative to unleaded gasoline. These
incentives as summarized in Table II include:

o The National Energy Act motor fuel excise tax exemption
   which applies through 1984 to gasoline/alcohol blends.
   This exemption is worth $.04 per gallon of gasoline/
   alcohol blend, or $0.40 per gallon or $16.80 per barrel
   of alcohol in 10 percent blends. The exemption does not
   apply to alcohol which uses petroleum, natural gas or
   coal as a feedstock. Some states have also exempted
   these blends from State excise taxes.

o Eligibility of alcohol fuels for Department of Energy
   entitlements, worth roughly 5 cents per gallon.

o Loan guarantees for alcohol pilot plants, administered
   through the U.S. Department of Agriculture, the
   Department of Housing and Urban Development, the Small
   Business Administration, the Economic Development
   Administration and Department of Energy.

o An additional 10 percent investment tax credit on top of
   the current 10 percent base. The value in Table II is
   based on a new 50 million gallon per year ethanol plant
   with a 20-year plant life and investment of $58 million
   (1978).

Table I. Biomass Feedstocks Potentially Available for Ethanol Fuel Production

BIOMASS FEEDSTOCK	MATERIAL POTENTIALLY AVAILABLE FOR ETHANOL PRODUCTION		PRODUCTION POTENTIAL, MILLION GALLONS PER YEAR ETHANOL
	MILLION DRY TONS	MILLION BUSHELS	
Cheese Whey	0.9	—	90
Corn	16.0	640	1,660
Grain Sorghum	2.7	110	280
Sugar Cane	2.6	—	150
Wheat	11.4	420	1,130
TOTAL			3,310

Table II. Impact of Incentives Cost of Alcohol Used in Gasohol

	SUBSIDY (Dollars)		EFFECTIVE NET PRICE (Dollars)	
	PER GALLON OF ALCOHOL	PER BARREL OF ALCOHOL	PER GALLON OF ALCOHOL	PER BARREL OF ALCOHOL
Base: Market Price (Dec. 1979) of ethanol used in gasohol			1.62	68.04
Subsidy:				
1. Exemption of Federal excise tax ($0.04 per gallon of fuel containing 10 percent alcohol	0.40	16.80	1.22	51.24
2. 20 percent investment tax credit	0.01	0.42	1.21	50.82
3. Entitlement credit (approximately $0.05 per gallon of ethanol)	0.05	2.10	1.16	48.72
4. State Incentive: Rebate of State Tax on Gasoline:				
Arkansas	0.95	39.90	0.21	8.82
California	0.50	21.00	0.66	27.72
Colorado	0.50	21.00	0.66	27.72
Connecticut	0.10	4.20	1.06	44.52
Indiana	0.32	13.44	0.84	35.28
Iowa	0.65	27.30	0.51	21.42
Kansas	0.50	21.00	0.66	27.72
Louisiana	0.80	33.60	0.36	15.12
Maryland	0.10	4.20	1.06	44.52
Montana	0.70	29.40	0.46	19.32
Nebraska	0.50	21.00	0.66	27.72
New Hampshire	0.50	21.00	0.66	27.72
North Dakota	0.40	16.80	0.76	31.92
Oklahoma	0.65	27.30	0.51	21.42
South Dakota	0.40	16.80	0.76	31.92
Wyoming	0.40	16.80	0.76	31.92

Net Energy Gain in Biomass Derived Ethanol Production.
Numerous studies have examined the net energy balance issue of
alcohol production and use. The majority conclude that the net
balance is small but positive, but exact estimates differ,
depending on the feedstock and process employed. It is expected
that advances in technology and ability to utilize feedstock
by-products more effectively will improve the net energy balance.
For example, a plant using food processing residue for feedstock
and coal for fuel may achieve a net reduction in imports
approaching its total production. Also, the use of coal in the
alcohol production process improves the oil savings attainable
through increased use of gasohol. Thus, the use of coal or
renewable fuels such as wood, agricultural residues, geothermal
energy or solar energy for alcohol production is highly
preferable to use of oil or natural gas.

Ethanol conversion facilities can readily be designed to use
fuel sources other than oil or natural gas. Small-scale on farm
plants can utilize corn stover as a boiler fuel, and larger
plants can rely on coal as a boiler fuel to produce process
steam. The ethanol production process can then be viewed as a
means of converting lower grade energy forms, such as coal,
distressed crops or solar energy, into premium transportation
fuel.

Environmental Effects of Ethanol Production. Production of
ethanol could potentially produce two forms of air pollution:
the release of pollutants from the boiler used to produce process
steam, and vaporization of ethanol during the production process.
If crop residues or lower grade fuels such as coal (low sulfur)
are used as boiler fuel, which is preferable from a net energy
gain basis, the resulting pollutants may be controlled through
use of flue gas stack scrubbers. The release of ethanol vapors
at the plant site is not considered a major concern at this time.

In small-scale farm production of ethanol a possible
environmental impact could occur through removal of crop residues
for use as a boiler fuel. Crop residues are important because
they help control soil erosion through their cover and provide
nutrients, minerals, and fibrous material which help maintain
soil quality. However, not more than one-third to one-half of
the residues from a grain crop devoted to ethanol production need
be used to fuel the process. Also, there are several methods,
such as crop rotation and winter cover crops, which lessen the
impact of crop residue removal.

A second environmental impact which might occur is related
to the application of thin stillage to the land. Thin stillage,
a product of the grain fermentation filtering process is composed
of very small solid particles and solubles. Two kinds of problems can result from applying thin stillage to the land: odor
and acidity. The impacts of applying thin stillage to the land

can be attenuated by using a sludge plow, possible recycling of
the thin stillage within the plant, or use of anaerobic digestion
to reduce the pollution potential of the thin stillage.

Technical Aspects of the Use of Ethanol-Gasoline Blends in Spark Ignition Engines

Consideration of the large scale use of ethanol-gasoline
blends in conventional gasoline spark ignition engines requires
evaluation of several key technical issues. In particular, in
blends with high proportions of ethanol, the characteristics and
performance may change substantially from the properties of gasoline. Evaluation of ethanol-gasoline blends, including gasohol
have focused on the following technical areas:
- o The relative differences in chemical and physical
  properties of ethanol, gasoline, and ethanol-gasoline
  blends
- o The performance of ethanol-gasoline blends and ethanol in
  conventional spark ignition engines
- o Effect of ethanol-gasoline blends on engine emissions
- o The compatibility of ethanol-gasoline blends with
  automotive fuel and engine systems

Chemical and Physical Properties of Ethanol and Gasoline.
The differing performance of ethanol and ethanol-gasoline blends
in conventional spark ignition engines compared to straight
gasoline can be attributed to the differences in chemical and
physical properties. While gasoline is a mixture of about 4 to
12 carbon atom hydrocarbons, ethanol is a single compound with
uniquely and narrowly defined properties. Differences in engine
performance and system compatibility between gasoline, ethanol,
and ethanol blends can be attributed predominantly to flash
point, boiling point, octane quality, heat of vaporization,
heating value, stoichiometric air/fuel ratio required for combustion, and water solubility. As a greater percentage of
ethanol is added to straight gasoline, the deviation of characteristics is approximately proportional to the percentage of
ethanol.
The octane-boosting properties of ethanol in gasoline were
particularly attractive at a time when higher octane lead-free
gasolines were in short supply and other octane enhancers such as
MMT (methylcyclopentadienyl manganese tricarbonyl) and lead are
under restrictions. For example, a three percent ethanol addition increases typical gasoline octane (measured as the average
of "research" and "motor" octane) by roughly one point. In a
gasohol mixture, the addition of 10 percent ethanol increases the
octane by 2 to 3 points depending on the composition of the
gasoline. Ethanol has been permitted by the Environmental
Protection Agency for use as a gasoline additive under Section

211(f) of the Clean Air Act. Two other chemicals - TBA (tertiary butyl alcohol) and MTBE (methyl tertiary butyl ether) also have been permitted as octane enhancers. However, both are currently made largely from petroleum.

Performance of Ethanol-Gasoline Blends in Spark Ignition Engines. A number of gasohol fleet tests have been run to evaluate performance under normal driving conditions. In addition, a number of authors have compiled technical information or performed laboratory and performance tests to evaluate various proportional blends of gasoline and ethanol (8,9,10). The gasohol fleet tests performed so far, have not been performed under the degree of rigorous controls and testing procedures that would ensure complete accuracy. However, none of the fleet testing which has been performed indicates any major problems with the use of gasohol in normal automotive use. The Department of Energy has received the results of extensive tests conducted by the state governments of Illinois, Nebraska and Iowa and also by the American Automobile Association. These tests indicate that the great majority of unmodified vehicles tested ran as well or better with gasohol than with fuel previously used. Specifically,

- o The American Automobile Association test showed that 88% of the vehicles ran as well or better on gasohol, which suggests that 12% of the vehicles showed negative results and would require some minor modification for comparable or improved performance on gasohol.

- o Illinois state government officials indicate that their test, (began in June 1978 and continuing beyond the date of this paper) which has used approximately 1,800 state vehicles including state police cars, has yielded positive results in each category tested.

- o The Iowa Development Commission (a state agency) conducted a 90 day gasohol marketing (customer opinion) test from June 15, 1978 to September 15, 1978, in which 232,000 gallons of gasohol were sold at five stations (also offering unleaded regular) across the state. Results show: a) 67% of users reported improved performance (29% cited increased mileage), b) three of every five users were repeat customers, c) 90% of users would purchase gasohol if it were available at most stations, d) gasohol outsold unleaded regular 3.9 to 1.

Gasohol is being used in automobiles without modification. With normal precautions to maintain a moisture free blend during refining/mixing, transportation, and at the service station, the tendency of the ethanol-gasoline mixture to undergo phase separation can be minimized. The presence of alcohol in the gasoline increases the water tolerance of the gasoline. While only a maximum of a few parts per million of water will mix freely with gasoline, a gasohol mixture will tolerate nearly 0.25 percent water (depending on temperature) before phase separation takes place. Because most automobiles have some small percentage of water in their gas tanks, the use of anhydrous ethanol will minimize phase separation problems.

Certain minor driveability problems can occur when using ethanol-gasoline blends in engines which are set-up for optimum performance with gasoline. With gasohol, under certain conditions, driveability problems may include:

o Greater tendency of stalling and stumbling during engine warm-up
o Increased tendency towards hesitation during acceleration for late model cars
o Increased vapor lock tendency in hot climates or at high altitude

Leaning of the air/fuel ratio in a gasoline engine is known to impair driveability. Driveability problems resulting from a lean air fuel mixture can include increased engine stalling and stumbling during engine warm-up, increased tendency to hesitate during acceleration, and increased tendency for engine surge during constant speed driving. The use of ethanol in automotive fuel leans the air/fuel mixture because the ethanol molecule contains oxygen, and increases the oxygen content in the engine combustion chamber. In late model vehicles which are adjusted to run lean in order to minimize pollution, the further leaning induced by use of blends can cause driveability problems. The same tendency is not typically present with older vehicles which have a richer initial air/fuel ratio.

A quantitative measure of warm-up driveability is presented in the test results of Figure 2. The results of these limited tests indicated that the use of a 10 percent by volume ethanol blend can be expected to degrade warm-up driveability of standard (i.e., non-adjusted and non-modified) carbureted cars to a degree ranging from negligible to pronounced (8). Further, at 22 percent ethanol, warm-up driveability was degraded significantly in all three cars tested. The three cars were tested by a procedure which measures driveability with a Total Weighted Demerit (TWD) value. (Higher values of TWD indicate poorer engine driveability). Carburetor modifications performed to achieve an air/fuel ratio for the blend which is equivalent to straight gasoline will result in improved driveability.

For engines which are adjusted or modified to operate at equivalent air/fuel ratios, the fuel economy differences between gasohol and gasoline are apparently negligible. However, the leaning effect of ethanol may result in fuel economy or exhaust emissions effects ranging from significant increases to significant decreases, depending on the adjustment of the engine. An analysis of state and private test data received by DOE indicates that use of a 90/10 blend of unleaded regular gasoline and ethanol (compared to 100% unleaded regular) resulted in similar

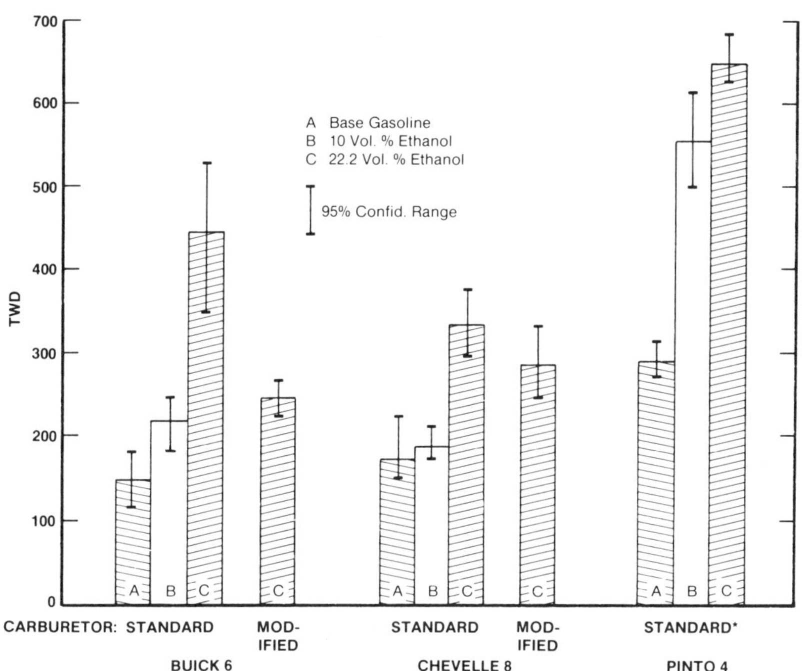

*Exhaust O_2 Sensor and Carburetor Feed-back Control

Figure 2. Warm-up driveability in three cars with gasoline and gasoline blends

miles per gallon or a small increase in mileage by a majority of
the vehicles tested. A substantial mileage increase or decrease
was found with a very few vehicles. Differing results in mileage
are attributable to the differing age, size, condition and
adjustment, weather conditions, and the differing quality of
ethanol and gasoline employed.

Effect of the Use of Ethanol Gasoline Blends on Engine
Emissions. On December 16, 1979, EPA approved use of gasohol
under Section 211(f)(3) of the Clean Air Act of 1977 and found
that there was no significant environmental risk associated with
the continued use of gasohol. Furthermore, new emissions control
systems, such as the "threeway catalyst with exhaust oxygen
sensors for carburetion feedback for air-fuel control," have been
shown to be equally effective using either gasoline or gasohol.

EPA and the Department of Energy have conducted a
cooperative gasohol testing program to obtain and evaluate
environmental impact data. On the basis of these tests, EPA
concluded that the addition of 10% ethanol to gasoline (11):
    o slightly decreases hydrocarbon emissions
    o significantly decreases carbon monoxide emissions
    o slightly increases nitrogen oxides emissions
    o substantially increases evaporative hydrocarbon emissions
The results to date have been generally favorable with
respect to the use of gasohol in automobiles. However, in a
recent technical memorandum, the Office of Technology Assessment
of the Congress of the United States stated that the "mixture of
observed emissions reductions and increases, and the lack of
extensive and controlled emissions testing, does not justify a
strong value judgment about the environmental effect of gasohol
used in the general automobile population (although the majority
of analysts have concluded that the net effect is unlikely to be
significant)"(12).

Compatibility of Ethanol-Gasoline Blends with Automotive
Fuel/Engine Systems. Experience with gasohol has indicated that
the solvent properties of ethanol loosen corrosion and dirt from
the walls of fuel tanks and fuel lines of automobiles. This
makes it advisable to flush and dry all storage tanks used with
ethanol-gasoline blends. Vehicle tanks, particularly with older
vehicles, should be flushed with ethanol or gasohol, and the fuel
filter may require replacement after the first or second tankful.
The use of neat ethanol or ethanol blends may potentially cause
minor problems with corrosion of metal fuel system materials,
particularly aluminum, copper, iron, lead and zinc. In addition,
clear polyamid, used in fuel systems for such items as fuel
filter housings, has been reported to fail in service with
ethanol blends.

Conclusions

Several considerations in the use of gasohol or ethanol-gasoline blends will ensure wide-scale application and maximum benefits to the United States:

- o Small and large-scale production of ethanol should be expanded utilizing existing food crop surpluses as well as distressed grains.
- o Ethanol plants should rely on lower utility fuel sources rather than premium fuels such as natural gas or petroleum derived fuels. Lower utility fuels may include coal, and renewable resources such as agricultural wastes (corn stover, bagasse) geothermal or solar energy.
- o Phase separation of ethanol-gasoline blends resulting from blend water content dictates the use of anhydrous ethanol and that the blend distribution system exercise additional care to minimize water contamination
- o The differences in fuel economy and exhaust emissions between a gasohol-type blend and straight gasoline are attributable to ethanol changing the effective air/fuel ratio of standard engines. Equal performance can be achieved by adjusting the air/fuel ratio to the equivalent value obtained with gasoline.
- o Research, development and testing should continue to:
  - Identify new feedstocks with high alcohol yields
  - Accelerate the development of feedstock collection and conversion technologies capable of using low-cost, waste feedstocks (including cellulosic materials).
  - Resolve the fuel consumption and exhaust emission impacts of ethanol-gasoline blends.
  - Remove the technical and material constraints to the use of alcohol fuels in all highway and non-highway systems including diesel engines, turbines and on-farm uses.

Literature Cited

1. Lucke, C. E.; and Woodward, S. M. "Use of Alcohol and
    Gasoline in Farm Engines," USDA Farmers Bulletin, 1907, 277.
2. "Fuels from Farms - A Guide to Small Scale Ethanol
    Production", Solar Energy Information Data Bank, for
    Department of Energy, February 1980.
3. Marks, C.. "The Fuel Quality and Quantity Requirements of
    Future Automobiles", General Motors Corp., NPRA AM-77-31,
    March 1977.
4. Balasubramaniam, M.; and Jenkins, R. "Automotive Fuel
    Requirements with Increasing Diesel Car Population and
    Implications to Refining Operation", TRW Energy Systems
    Planning Division, McLean, Virginia, PA-3885-22-1, February
    1978.
5. "The Report of the Alcohol Fuels Policy Review", U.S.
    Department of Energy, Washington, D.C., June 1979.
6. "Grain Motor Fuel Alcohol Technical and Economic Assessment
    Study", Raphael Katzen Associates for U.S. Department of
    Energy, June 1979.
7. "Ethanol Production from Biomass with Emphasis on Corn",
    College of Agriculture and Life Sciences, University of
    Wisconsin, Madison, September 1979.
8. Nakaguchi, G. M.; and Keller, J. L. "Ethanol Fuel
    Modification for Highway Use", Final Report, Union Oil
    Company of California, for U.S. Department of Energy, 26
    July 1979.
9. Ecklund, E. E. "A Department of Energy View of Gasohol",
    Speech Delivered to American Gasohol Limited, New York,
    1979.
10. Bernhardt, I. W.; Lee, W., "Possibilities for Cost-Effective
    Use of Alcohol Fuels in Otto Engine-Powered Vehicles",
    Proceedings, International Symposium on Alcohol Fuel
    Technology, Methanol and Ethanol, November 21-23, 1977.
11. "Analysis of Gasohol Fleet Data to Characterize the Impact
    of Gasohol on Tailpipe and Evaporative Emissions", U.S.
    Environmental Protection Agency, Technical Support Branch,
    Mobile Source Enforcement Division, December 1979.
12. "Gasohol--A Technical Memorandum", Office of Technology
    Assessment, September 1979, 69 p., Available from
    Superintendent of Documents,
    U.S. Government Printing Office, Stock No. 052-003-00706-1.

RECEIVED February 23, 1981.

Manufacture of *n*-Butanol and 2-Ethylhexanol by the Rhodium Oxo Process and Applications of the Alcohols

C. E. O'ROURKE, P. R. KAVASMANECK, and R. E. UHL

Union Carbide Corporation, Danbury, CT 06817

The oxo process consists of the reaction between an olefin, carbon monoxide, and hydrogen, in the presence of a catalyst (rhodium or cobalt) to form an aldehyde with one more carbon atom than the starting olefin.

$$R-CH=CH_2 \; + \; CO \; + \; H_2 \; \xrightarrow{\text{Catalyst}} \; R-CH_2-CH_2CHO$$

The only exception is with the cobalt catalyst when the starting olefin is ethylene, in which case both the aldehyde and diethyl ketone are formed.

The discovery of the oxo (or more accurately, hydroformylation) process was by Ruhrchemie's Roelen who applied for a German patent in 1938. This turned out to be a breakthrough method for both higher and lower alcohols production from the aldehydes. Exploitation of the process, however, was delayed by World War II. It was not until 1948 that commercialization occurred with the manufacture of isooctyl alcohol by Enjay Chemical Company (1).

Subsequently, a whole host of both lower (C_3, C_4, C_5) and higher (C_6, C_7, C_9, C_{10}, C_{11}, etc.) oxo alcohols have been commercialized. Of all these alcohols, the most important by far have turned out to be n-butanol and 2-ethylhexanol - both of which are derived from n-butyraldehyde based on hydroformylation of propylene. In addition to n-butyraldehyde, the lower valued isobutyraldehyde is produced as a by-product. Some of this is converted to isobutanol.

Propylene Oxonation Processes

There are three commercially significant oxo processes starting with propylene for the production of butanols and 2-ethylhexanol. The primary difference in these processes is associated with the catalyst system used to produce butyraldehyde. The catalysts are:

<u>TABLE I</u>

<u>Comparison of Processes for Propylene Oxonation</u>

	Conventional Cobalt	Cobalt Ligand	Rhodium Ligand
Operating Pressure	High	Medium	Low
Operating Temperature	Medium	Medium	Low
Catalyst Cycle	Complex	Complex	Simple
n/i Isomer Ratio	3/1	9/1	$\geq$10/1
Primary Product	Aldehydes	Alcohols	Aldehydes
Other By-Products	High	Medium	Low

o   conventional cobalt
o   cobalt ligand
o   rhodium ligand

Each catalyst system employs a unique set of operating conditions. A brief comparison of the three processes is shown in Table I.

With the conventional cobalt carbonyl catalyst, the ratio of n- to isobutyraldehyde produced is between 3 and 4:1. This low ratio represents a significant loss of propylene and synthesis gas (20-25%) to the lower valued or unwanted isobutyraldehyde. In lieu of having large and expensive storage facilities, isobutyraldehyde or isobutanol is frequently burned for fuel value. An alternative to burning is the expensive catalytic cracking of isobutyraldehyde to form synthesis gas and propylene for recycle and consumption in the hydroformylation step (2).

Considerable effort over the years has been devoted to a search for new oxo catalysts. This has been motivated by a desire to minimize the less valuable isobutyraldehyde/alcohol and also to lower oxo reaction temperatures and the high pressures (3-4000 psi) associated with the conventional cobalt process for reduced capital investment and increased energy savings.

In the case of the cobalt ligand process, alcohols rather than aldehydes are the main products. While the cobalt ligand catalyst overcomes the isomer ratio disadvantage, it has only one-fifth the activity of the conventional cobalt catalyst. Even at temperatures of 180-200°C, space time yields remain lower than in the conventional cobalt process. At the elevated temperatures the cobalt ligand catalyst is a highly active hydrogenation catalyst that converts a significant portion (10%) of the propylene to propane and most of the butyraldehyde to butanol.

The rhodium ligand process, which will be discussed in greater detail in the following sections, operates at lower pressures and temperatures and overcomes the high by-product levels as well as the isomer ratio disadvantage of the conventional cobalt catalyst.

Rhodium Low Pressure Oxonation

Union Carbide, Johnson Matthey and Davy McKee have jointly developed a low pressure, two-step rhodium catalyzed process which takes place at around 110°C and 100-300 psi (3, 4). The rhodium complex $HRh(CO)(P\phi_3)_3$ is stable at typical reaction temperatures and pressures. The rhodium concentration in the reaction mix is usually around 0.02 wt%.

A summary of the key features of this rhodium catalyzed low pressure oxo (LPO) process is as follows:

o   <u>Low Pressure Operation</u> - Use of a highly active
     rhodium catalyst allows operation at pressures of
     300 psi and less, far lower than with cobalt catalyst.

o   <u>Efficient Feedstock Utilization</u> - Aldehyde normal to
     iso ratios of 10:1 or greater are achieved with
     consequent reduction in losses of propylene and
     synthesis gas to isobutyraldehyde. In addition,
     losses to propane are reduced to 2 percent or less and
     losses to heavy by-products are insignificant.
     Hydrogenation of butyraldehyde to butanol is virtually
     nil, leading to a simpler product mix and the absence
     of acetals.

o   <u>Lower Temperature</u> - The LPO process operates 45-85°C
     lower than cobalt oxo processes. This significantly
     reduces efficiency losses from trimerization of
     butyraldehyde and from other side reactions.

o   <u>Simplicity</u> - The cumbersome and corrosive catalyst
     recycle systems of many conventional cobalt oxo
     processes are avoided.

o   <u>High Catalyst Productivity</u> - The high catalyst
     productivity leads to a low rhodium inventory and low
     catalyst cost per pound of product.

o   <u>Low Energy Requirements</u> - The low process operating
     pressure means that synthesis gas compression can be
     avoided. The process involves little distillation for
     product or by-product, thereby minimizing steam
     requirements.

High concentrations of CO reduce alcohol and paraffin formation
but lower the n/i ratio. Excess ligand, i.e., $P\phi_3$ that is
not chemically bound to Rh, is maintained at around 5-8 wt% in
the reaction mix to achieve a high ratio of normal to iso
aldehydes.

<u>Catalyst Description</u>. The LPO catalyst is a
triphenylphosphine modified carbonyl complex of rhodium.
Triphenylphosphine, carbon monoxide, and hydrogen form labile
bonds with rhodium. Exotic catalyst synthesis and complicated
catalyst handling steps are avoided since the desired rhodium
complex forms under reaction conditions. Early work showed
that a variety of rhodium compounds might be charged initially
to produce the catalyst. Final selection was made on the basis
of high yield of the catalyst precursor from a commodity
rhodium salt, low toxicity, and good stability to air, heat,
light, and shock.

The exact reaction mechanism is uncertain, but it is believed to be as shown in Figure 1. The proposed mechanism illustrates factors that provide selectivity to normal product.

The initial step is coordination of propylene with species A. Propylene complex B rearranges to the alkyl complex C, which undergoes CO insertion to form the acyl derivative D. Oxidative addition of hydrogen gives the dihydroacyl complex E. Finally, hydrogen is transferred to the acyl group. Butyraldehyde is formed along with the 4-coordinate species, F.

The presence of an excess of triphenylphosphine ligand L under the low pressure conditions favors high selectivity to normal aldehyde. The excess ligand suppresses dissociation of species A into one containing only a single phosphine ligand. By favoring the presence of species A, when olefin attacks, the steric effect of the two bulky triphenylphosphine ligands favors a high ratio of primary alkyl. If few triphenylphosphine ligands were present in the complex, a higher proportion of propylene would form secondary alkyl groups, giving more isoaldehyde.

n-Butanol Manufacture Using a Rhodium Catalyst for The Oxo Reaction

A simplified flow diagram of the rhodium catalyzed oxo process is shown in Figure 2 (3, 4, 5). The synthesis gas and propylene streams are first purified via proprietary solid-absorbent techniques to remove catalyst poisons such as hydrogen sulfide or carbonyl sulfide. Then they join with recycle gas, and the combined stream enters the base of the hydroformylation reactor through a distributor. Here, the rhodium-based catalyst is present in a homogeneous liquid phase, dissolved together with free triphenylphosphine in a mixture of butyraldehyde and heavy by-products from aldehyde trimerization.

These aldehyde-condensation products are solvents for the catalyst. This allows the unit to operate without need for other solvents and forestalls the need for a catalyst-recycle step.

Because of the high activity of the catalyst, only a low rhodium concentration of several hundred parts-per-million is needed. The triphenylphosphine level is kept much higher-typically several percent by weight.

Reaction temperatures typically are 80 to 120°C at pressures between 200 and 300 psi. Heat of reaction is taken out partly via vaporization of aldehydes into the overhead gas stream, and partly by circulating a coolant through coils inside the reactor.

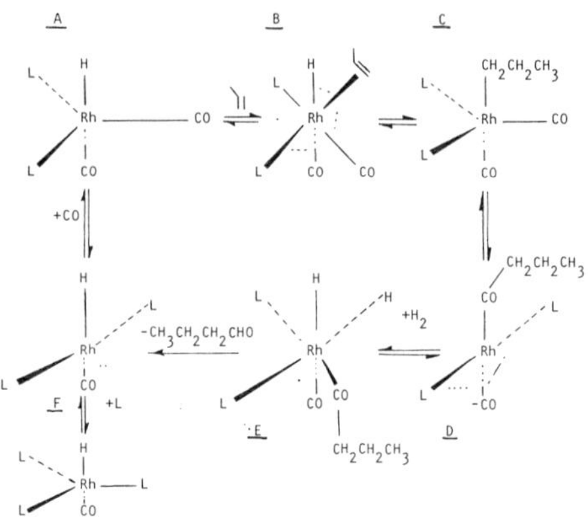

Figure 1.   Possible catalyst mechanism

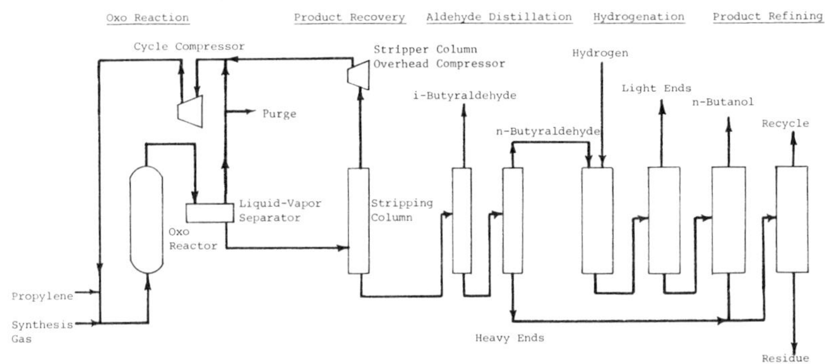

Figure 2.   n-Butanol manufacture using a rhodium catalyst for the oxo reaction (3, 5)

Reactor effluent is removed in the gaseous phase, passing through entrainment separators that minimize rhodium loss. This effluent then goes to a condenser where aldehydes and by-products drop out; this mixture is removed in a separator. The liquid stream from the separator contains appreciable amounts of dissolved gases, mainly propylene and propane. A product stripping column distills these out. The liquid stream from this stripper goes through two distillation columns in series that remove iso- and n-butyraldehyde as overhead products, respectively. A small stream that contains heavy by-products formed in the reactor leaves the bottom of the second column. This stream can be combined with the heavy ends stream from the n-butanol column and valuable aldehydes and alcohols recovered for recycle. The iso-butyraldehyde overhead product from the first aldehyde column may be hydrogenated and sold as a low cost solvent, cracked to synthesis gas and recycled to the oxo reactors, or burned as fuel.

The propylene-rich overhead product from the stripping column is compressed and recycled into the gas stream to the reactor. The overhead gas from the liquid-vapor separator likewise is compressed and recycled after a portion of it is bled off in a purge stream that controls the level of propane and other inert gases introduced with the feedstock in the gas circuit.

<u>Butyraldehyde Hydrogenation and Refining</u>. A wide variety of hydrogenation processes using copper, nickel or combined copper-nickel, fixed bed catalysts in either the vapor or liquid phase are used commercially (<u>6</u>, <u>7</u>, <u>8</u>, <u>9</u>). After the hydrogenation step the butanol is refined using a two-column scheme. The first column removes water and a small amount of light ends produced during hydrogenation. The second column takes refined n-butanol as an overhead product. The small heavy ends stream from the second column is combined with the heavy ends stream from the n-butyraldehyde column for recovery and recycle of contained alcohols and aldehydes. The final residue stream is burned as fuel.

2-Ethylhexanol Manufacture Using A Rhodium Catalyst For The Oxo Reaction

The production of 2-ethylhexanol from propylene by the rhodium catalyzed, low pressure oxo process is accomplished in three chemical steps. The first step of the process (described in section on n-butanol manufacture) converts propylene to normal butyraldehyde by hydroformylation in the presence of a rhodium catalyst. In a second step, the normal aldehyde is aldoled to form 2-ethylhexenal. 2-Ethylhexenal is then hydrogenated to 2-ethylhexanol and refined in the third and final step (see Figure 3).

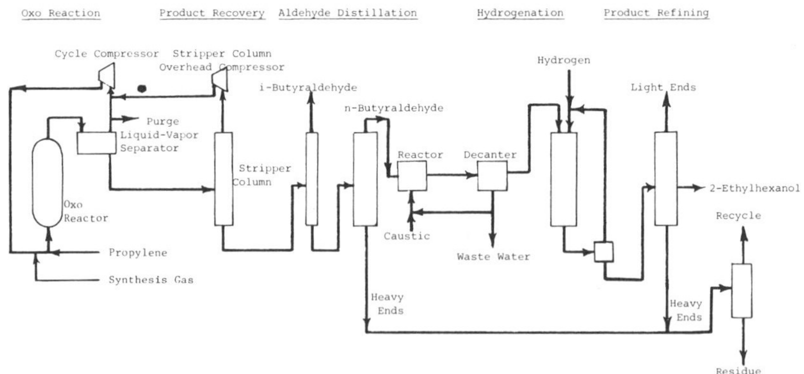

Figure 3.   2-Ethylhexanol manufacture using a rhodium catalyst for the oxo reaction (3, 5)

Aldolization, Hydrogenation and Refining. In the second
processing step, aldolization (10), n-butyraldehyde is added to
aqueous sodium hydroxide which is recycled from the decanter.
The caustic concentration in the cycle is maintained at 3% by
adding makeup caustic. Residence time in the aldol reactor is
typically about 60 seconds. Reaction temperature is controlled
between 110°C and 120°C at pressures of 50 to 70 psi. The
reactor effluent is fed to a decanter to separate the
2-ethylhexenal (EPA) from the caustic solution recycle. Water
produced by dehydration of the aldol dilutes the cycle and is
purged after decantation.

The third processing step receives the organic phase from
the decanter for hydrogenation to 2-ethylhexanol. As in the
case of butanols, the hydrogenation of EPA is accomplished by a
wide variety of processes (6, 7, 8, 9). In the process shown
in the flow diagram, EPA is vaporized into a recycle hydrogen
stream and reacted over a copper or copper-nickel catalyst at
150° to 200°C at 50 to 70 psi. Crude 2-ethylhexanol is
condensed from the hydrogen cycle and refined in a single
column which takes water and light ends overhead and heavy ends
as a tails stream. Refined 2-ethylhexanol is taken as a side
stream make. Heavy ends from the n-butyraldehyde column and
the 2EH column are fed to a second column to recover and
recycle aldehydes and alcohols. The residue is burned as fuel.

Applications

The key industrial applications and markets for normal and
isobutanol and 2-ethylhexanol are discussed . As will be
noted, the C_4 oxo alcohols find use primarily within the
coatings industry, either as solvents, per se, or as
intermediates to manufacture solvents or protective coatings
chemicals. Applications for 2-ethylhexanol, while numerous and
varied, are basically oriented toward the manufacture of
plasticizers for polyvinyl chloride. Total U.S. consumption of
these alcohols in 1979 was approximately 1.3 billion pounds -
730 million pounds of n-butanol, 190 million pounds of
isobutanol, and 380 million pounds of 2-ethylhexanol. The
consumption pattern is summarized in Table II and described in
the following sections:

n-Butanol. This alcohol is widely used as a solvent or
co-solvent in industrial protective coatings formulations. It
has traditionally been the preferred solvent or co-solvent in
applications where slower drying is desired. The major
segments of the n-butanol coatings solvent market are:

o   nitrocellulose lacquers used as both top coats and
        sealers in wood finishing applications.

<u>Table II</u>

<u>n-Butanol, Isobutanol, 2-Ethylhexanol</u>
<u>Consumption Pattern - U.S.A.</u>

End Use	Consumption Pattern, 1979 % of Total
<u>n-Butanol</u>	
Butyl Acrylate	27
Coatings Solvent	22
Butyl Glycol Ethers	22
Butyl Acetate	13
Plasticizers	8
Butyl Amines	2
Butyl Methacrylate	2
Lube Oil Additives	1
Miscellaneous	3
	100
<u>Isobutanol</u>	
Isobutylamines	32
Coatings Solvent	21
Isobutyl Acetate	16
Lube Oil Additives	13
Plasticizers	8
Miscellaneous	10
	100
<u>2-Ethylhexanol</u>	
Plasticizers	72
Acrylates	15
Oil and Fuel Additives	6
Surfactants	4
Miscellaneous	3
	100

o   thermoset acrylic resin coating systems applied mainly
     in the metal furniture field, appliances, and coil
     coatings.

o   acrylic-melamine resin coating systems (70%
     acrylic/30% melamine) used in the automotive
     refinishing and top coat markets where ease of
     application and low cost are prime considerations.

o   amino resin systems such as alkyd-urea and
     alkyd-melamine.

o   water-reducible coatings (as a co-solvent).

Coatings solvent applications account for approximately 22
percent of total n-butanol demand in the United States. The
market is predicted to remain relatively stagnant over the next
five years and then gradually decline as water-borne and other
higher solids coatings systems are developed in response to the
legislative and various socio-economic forces facing the
coatings industry.

n-Butyl acrylate, made by esterification of the alcohol
with acrylic acid, is a major application representing about 27
percent of n-butanol domestic demand. Acrylic coating resins
based on n-butyl acrylate, ethyl acrylate, and methyl
methacrylate established their position in the industrial
coatings market during the early 1960's. These systems are
widely used in auto top coats and coil and appliance coating
applications. In the trade paint area, exterior house paints
based on water-borne acrylic latexes have largely displaced
oil-based alkyd systems in this market. n-Butyl acrylate is
expected to grow at a rate of about 6 percent per year over the
next five years.

Also in the acrylate family, n-butyl methacrylate is
produced by the same process as methyl methacrylate -
alcoholysis of methacrylamide with n-butanol. There are two
major applications for n-butyl methacrylate - as a modifier in
polymers for industrial and automotive lacquers and as a
comonomer with higher methacrylates (e.g., lauryl and stearyl)
in the manufacture of lubricant viscosity index improvers.

The butyl glycol ethers are also primarily coatings end use
oriented. Approximately 22 percent of U.S. n-butanol demand is
consumed in the manufacture of ethylene glycol mono-n-butyl
ether (butyl CELLOSOLVE[R]) and diethylene glycol mono-n-butyl
ether (butyl CARBITOL[R]).

The principal application for butyl CELLOSOLVE is as a low
evaporating solvent of choice in many industrial coatings
applications including coil coating and wood finishing. Its
acetic acid ester, butyl CELLOSOLVE acetate, finds use in
similar applications as a coalescing aid in latex trade

paints. As an ingredient in many specialty household products,
butyl CELLOSOLVE (and other glycol ethers) has three main
functions:

-   as an aid in removal of water soluble or water
    dispersible soils.
-   as a coupling agent.
-   as a filming or coalescing aid in water-borne floor
    polishes.

Butyl CARBITOL is used in trade paint formulations and, as in
the case of its acetic acid ester, as a coalescing aid. Small
amounts of triethylene glycol mono-n-butyl ether are used in
hydraulic brake fluid systems. The butyl glycol ethers are
increasing in volume at a rate of about 5 percent per year.

Demand for n-butanol as n-butyl acetate accounts for about
13 percent of domestic consumption of the alcohol. This
acetate ester is a general purpose coatings solvent with
particular application in nitrocellulose lacquers and moisture
cured urethane coatings formulations. n-Butyl acetate is the
traditional medium boiling solvent for nitrocellulose lacquers
and, as such, is the standard by which other solvents are
compared.

Manufacture of butylamines (n-butanol and ammonia) accounts
for about 2 percent of n-butanol demand. The C_4 amines are
used primarily as intermediates in the manufacture of
herbicides, rubber chemicals, and pharmaceuticals. Consumption
of n-butanol in butylamines is growing at a rate of about 5
percent per year.

Use of n-butanol in the manufacture of a wide range of
plasticizers for PVC and/or other coatings resins represents
about 8 percent of domestic demand for alcohol. The
principal end use products involved include:

    dibutyl phthalate
    butyl benzyl phthalate
    dibutyl maleate (comonomer for internal plasticization)
    dibutyl fumarate (comonomer for internal plasticization)
    dibutyl sebacate
    butyl stearate

Butyl benzyl phthalate (BBP) is by far the largest volume
outlet for n-butanol in the plasticizer market. Monsanto is
the world's principal supplier of this fast fusing plasticizer
for PVC resin. BBP has become a key plasticizer in vinyl
flooring compositions as a result of its stain resistance and
processability.

Miscellaneous uses for n-butanol including use as an
intermediate in the manufacture of lube oil additives and
pharmaceuticals account for about 4 percent of U.S. demand.

Isobutanol. Isobutanol is consumed primarily as a solvent in surface coatings applications, either directly, or as isobutyl acetate. Both the alcohol and its acetic acid ester are substitutes for n-butanol and n-butyl acetate in many coatings uses, but a significant price differential would be required for major replacement to occur. The iso products are, as a rule, faster evaporating and have somewhat poorer solvency relative to their normal counterparts, hence the normals are preferred and command higher prices. Abnormally high usage of isobutanol occurred in the solvents market in 1979 due primarily to a shortage of n-butanol. In all, about 37 percent of domestic isobutanol demand is centered in the coatings solvent market, as the alcohol per se, or as the acetic acid ester.

Diisobutylamine production (isobutanol and ammonia) for use in the manufacture of herbicides is the largest single chemical intermediate market for isobutanol. It represented approximately 32 percent of U.S. demand for the alcohol in 1979. This application is growing at a rate of roughly 3 percent per year.

Isobutanol use in the manufacture of zinc dialkyl dithiophosphates (ZDDP), anti-wear lube oil additives, represented 13 percent of domestic consumption. Other alcohols used in this application include methylamyl alcohol, primary amyl alcohol, n-butanol, 2-ethylhexanol and isooctanol.

The only remaining use of any importance is the manufacture of a number of specialty phthalate plasticizers for PVC and other resin systems. Plasticizers account for about 8 percent of demand and the remaining 10 percent involve a variety of miscellaneous applications.

2-Ethylhexanol. Use of 2-ethylhexanol in the manufacture of PVC plasticizers, most notably DOP (di-2-ethylhexyl phthalate), has historically accounted for over 70 percent of U.S. demand for this alcohol. DOP has been the most widely used general purpose PVC plasticizer for close to half a century and is considered the "workhorse" of the industry. The material is used in a wide variety of PVC resin applications including flooring, wire and cable, packaging and coated fabrics. In the past, DOP has represented the industry cost/performance standard against which all competing plasticizers were measured.

Shutdown of Oxochem's 2-ethylhexanol unit in 1978 eliminated 300 million pounds of capacity overnight. As a result, DOP users began a crash reformulation effort on alternative plasticizers. Linear alcohol and C_9 and C_{10} branched chain alcohol based phthalates as well as butyl benzyl phthalate, for example, were found to be wholly or partially substitutable plasticizers in many DOP applications. New 2-ethylhexanol capacity available starting in 1981 is expected

to stem the decline in DOP demand and, although the product
will probably not regain its dominant position in the industry,
further growth should track flexible PVC demand of about 3
percent per year.

Other 2-ethylhexanol based plasticizers include
di-2-ethylhexyl adipate for vinyl food wrap application,
tri-2-ethylhexyl trimellitate, di-2-ethylhexyl azelate, and
di-2-ethylhexyl sebacate.

2-Ethylhexyl acrylate manufacture represented about 15
percent of domestic consumption of the alcohol. The acrylate
is the longest chain acrylate ester produced by esterification
of acrylic acid. The monomer is used in acrylic copolymers for
pressure sensitive adhesives, PVC impact modifiers, and as a
comonomer with vinyl acetate and vinyl chloride in latexes for
paints and textiles. Growth over the next 5 years is estimated
at 6 percent per year.

The alcohol also finds use in the manufacture of lube and
fuel oil additives and synthetic lubricants (about 6 percent of
domestic consumption). The zinc dialkyl dithiophosphate
anti-wear additive based on 2-ethylhexanol provides ideal
compatability, oil solubility, and high temperature stability
in many lube oils for both spark ignition and diesel engines.

In a fuel additive use for 2-ethylhexanol, the alcohol is
reacted with nitric acid and the nitrate produced added in low
concentrations to diesel fuel to improve ignition quality. Use
of such a "cetane improver" allows the refiner greater
flexibility in blending distillate into fuel such as aromatics,
branched paraffins, and naphthenes. The "cetane improver" also
helps meet sudden local increases in demand for diesel fuels by
allowing refiners to upgrade stored heating oil to diesel fuel
quality quickly and economically.

Diester synthetic base stock lubricants formulated with
2-ethylhexanol (e.g., di-2-ethylhexyl adipate) provide
excellent low temperature starting properties in automotive
crank case applications and are also employed as lubricants for
industrial machinery such as compressors and turbines.

About 4 percent of 2-ethylhexanol domestic demand is
consumed in the manufacture of di-2-ethylhexyl sulfosuccinate -
a general purpose anionic surfactant used in textile processing.

Miscellaneous uses for 2-ethylhexanol include specialty
surfactants, anti-foaming agents, esters of boric and
chloroformic acid, and specialty solvents.

Literature Cited

(1) Hatch, L.F., "Higher Oxo Alcohols", Enjay Chemical Co.,
     New York, 1957, p. 1.

(2) Falbe, J., Tummes, H., and Hahn, H. D.; U. S. Pat.
     4,039,584 (Aug. 2, 1977)

(3) Chemical Engineering, 1977, <u>48</u>, 26.

(4) Pruett, R. L., Smith, J. A., Brit. Patent 1,338,237.
     (Union Carbide).

(5) Anon., Hydrocarbon Processing, November, 1975, 54, 117.

(6) Hull, L. E. C., US Patent 3,288,866, November 29, 1966.
     (to Distillers).

(7) Hull, L. E. C., Pickup, K. G., US Patent 3,431,311,
     March 4, 1969. (to Distillers).

(8) Grinstein, R. H., US Patent 3,646,227; February 23,
     1972. (to Shell Oil Company).

(9) Diekhaus, G., Liethen, O., German Patent 2,047,437,
     May 10, 1972. (to Ruhrchemie).

(10) Brit. Patent 1,462,328; January 26, 1977. (to
     Ruhrchemie).

RECEIVED March 13, 1981.

Manufacture of Higher Straight-Chain Alcohols by the Ethylene Chain Growth Process

PAUL H. WASHECHECK

Conoco Chemicals, Conoco Inc., P.O. Drawer 1267, Ponca City, OK 74601

The first commercial volumes of higher molecular weight, linear alcohols were prepared from naturally occurring fats and oils. Because of this origin, these alcohols, even the synthetic versions, are commonly called "fatty alcohols." Another common name for these products is "detergent-range alcohols," after their highest volume use. This name helps to distinguish alcohols with eleven or more carbons from alcohols with six to ten carbons which are known as "plasticizer alcohols," after their highest volume use.

For this discussion we will arbitrarily define higher molecular weight or detergent-range alcohols as those with eleven or more carbon atoms. As with many boundaries, the one between plasticizer and detergent alcohols is not distinct. Many detergent intermediates contain only ten carbon alcohols. and Shell has recently introduced a new product which contains the nine through eleven carbon homologs. On the other hand, isotridecyl alcohol, a C_{13} oxo alcohol, is commonly used as a plasticizer feedstock.

Coconut oil and tallow were the principal raw material sources for early fatty alcohol manufacture. Coconut oil is a lauryl-range oil and affords primarily C_{12} and C_{14} alcohols. Tallow is a stearyl-range oil and yields primarily C_{16} and C_{18} alcohols. Both of these natural products form only even carbon-numbered alcohols. Some synthetic alcohols contain both even and odd carbon-numbered alcohols while other synthetic alcohols are like the natural products and contain only even carbon-numbered homologs.

It was recognized in the late 1950's that the supply and therefore price of naturally occurring products, such as coconut oil and tallow, could fluctuate widely. Both political and climatic conditions have an influence on the supply and price of these oils, and the price of naturally-derived alcohols depends on their raw material costs. With both politics and climates to content with, natural alcohol producers often have little control over the price of their final product.

This situation provided an incentive for development of synthetic routes to fatty alcohols.

Synthetic fatty alcohols fall into three broad categories and are manufactured from two basic raw materials--ethylene and n-paraffins. One group is secondary alcohols which are prepared by oxidation of n-paraffins in the presence of boric acid. A second group consists of oxo alcohols manufactured by hydroformylation of linear olefins which are derived from either n-paraffins or ethylene. Both of these alcohol types are discussed in separate chapters. The last group is Ziegler alcohols which are prepared from ethylene and are the primary subject of this chapter.

Secondary alcohols are much different chemically than primary alcohols, such as natural alcohols. In addition, commercial secondary alcohols are prepared from both even and odd carbon-numbered n-paraffins and thus contain both even and odd carbon-numbered alcohols. Oxo alcohols are primary alcohols, as are natural alcohols. However, oxo alcohols contain from twenty to sixty percent branched chain alcohols and also contain both even and odd carbon-numbered homologs. Ziegler alcohols are very similar to natural alcohols. They are primary alcohols and are a mixture of only even carbon-numbered homologs. The major differences between Ziegler and natural alcohols are trace impurities present and the range of synthetic products, C_4-C_{30}, available.

Natural Alcohols

As noted earlier, natural alcohols are produced from coconut oil and tallow as well as some other fats and oils--palm kernel oil, palm oil, sperm whale oil, etc. Most of these natural oils actually consist of fatty triglycerides, i.e., glycerol esterified by three molecules of fatty acid. There is very little free alcohol present in these materials, and the alcohols are derived from the fatty acid moiety of the triglyceride by reduction.

Each triglyceride molecule has a random distribution of acid chain lengths and degrees of unsaturation. However, the composition of fats and oils from a common source is relatively uniform. While fatty alcohols theoretically could be derived from any fat or oil, most are prepared from coconut oil or tallow with an increasing quantity derived from palm oil or palm kernel oil. Approximate compositions (1) of these four oils are listed in Table I.

The first commercial production of fatty alcohols in the 1930's employed a sodium reduction process (Bouveault-Blanc) (2). However, the high usage (4 mol/mol alcohol) of expensive sodium soon led to replacement of this method of reduction by catalytic hydrogenation.

TABLE I

Fatty Acid Composition, Weight Percent

	Coconut Oil	Palm Kernel Oil	Tallow	Palm Oil
C_6	0.8	–	–	–
C_8	5.4	2.7	–	–
C_{10}	8.4	7.0	–	–
C_{12}	45.4	46.9	–	–
C_{14}	18.0	14.1	6.3	1.4
C_{16}	10.9	8.8	27.4	40.1
C_{18}	9.9	20.5	66.2	58.5
C_{20}	0.4	–	–	–

There are basically two types of hydrogenation processes--hydrogenation of methyl esters and hydrogenation of fatty acids--with the former predominating in the U.S. Ashland and Procter & Gamble, the two U.S. fatty alcohol producers, both utilize methyl ester-based processes ($\underline{3}$).

In the methyl ester route (Figure 1), refined triglycerides are reacted with methanol in the presence of a sodium methoxide catalyst to form the corresponding fatty acid methyl esters and glycerine ($\underline{2}$).

$$C_3H_5(OCR)_3 + 3CH_3OH \xrightarrow{\text{NaOMe}} 3R\overset{O}{\overset{\|}{C}}OCH_3 + C_3H_5(OH)_3$$

This is a batch reaction carried out at atmospheric pressure with refluxing methanol. The crude reaction mixture is phase-split, and commercial glycerine is recovered from the lower phase by separating excess methanol.

The methyl esters can be fed directly to a hydrogenation unit, or more often are distilled to separate unreacted triglycerides which are recycled. The distilled methyl esters are hydrogenated at approximately 300°C and 3000 psig with a copper chromite catalyst in slurry form through tubular reactors. Excess hydrogen is used for reduction as well as agitation of the slurried catalyst.

$$R\overset{O}{\overset{\|}{C}}OCH_3 + 2H_2 \longrightarrow RCH_2OH + CH_3OH$$

The product mixture is flashed to separate hydrogen and some methanol. Catalyst is then separated from crude alcohol via centrifugation and recycled to the reactor as a slurry.

The last traces of catalyst are removed from crude alcohol by filtration. Crude alcohol is then stripped at atmospheric pressure to remove methanol and vacuum distilled to isolate the desired alcohol fractions.

Ziegler Alcohols

In the mid 1950's, Dr. Karl Ziegler ($\underline{4}$) and his associates at the Max Planck Institute carried out fundamental research which provided the basis for schemes to synthesize even carbon-numbered, linear alcohols similar to natural alcohols. Commercial plants were built by Conoco in 1962, Condea in 1964, and Ethyl in 1965. The Conoco and Condea processes are virtually identical and different from the Ethyl process.

There are four basic parts of a Ziegler alcohol process.
1. Preparation of triethylaluminum.
2. Growth with ethylene to higher alkylaluminum compounds.
3. Oxidation to aluminum alkoxides.
4. Hydrolysis to product alcohols.

Both the Conoco and Ethyl processes use these four basic steps, and Ethyl adds a transalkylation step to control product distribution.

Synthesis of triethylaluminum from aluminum, hydrogen, and ethylene is the first segment of a Ziegler alcohol process. It can be carried out in a single step, but normally is accomplished on a commercial scale in two stages with recycle of two-thirds of the trialkylaluminum product.

$$4Al(C_2H_5)_3 + 2Al + 3H_2 \longrightarrow 6Al(C_2H_5)_2H$$

$$6Al(C_2H_5)_2H + 6CH_2{=}CH_2 \longrightarrow 6Al(C_2H_5)_3$$

$$2Al + 3H_2 + 6CH_2{=}CH_2 \longrightarrow 2Al(C_2H_5)_3$$

Hydroaluminum of triethylaluminum forms diethylaluminum hydride. A Conoco patent ($\underline{5}$) indicates the hydrogenation reaction is catalyzed by titanium and zirconium in the feed aluminum and probably does not occur with ultrapure aluminum. Ethylene readily adds to the resulting diethylaluminum hydride to yield triethylaluminum.

In the chain growth step, ethylene adds to triethylaluminum to form higher trialkylaluminum compounds with an even number of carbon atoms.

$$Al(C_2H_5)_3 + mCH_2{=}CH_2 \longrightarrow Al[(CH_2CH_2)_mCH_2CH_3]_3$$

The mole percent distribution of alkyl chains follows a Poisson curve ($\underline{6},\underline{6a}$). The average number of ethylene units added during this growth step is commonly referred to as the "m-value" and is a good description of the distribution of alkyl groups on aluminum.

Addition of ethylene to triethylaluminum is first-order in monomeric triethylaluminum and ethylene (7). Since triethylaluminum is largely dimeric in the liquid phase, kinetics of the growth reaction take the following form:

$$[Al(C_2H_5)_3]_2 \rightleftharpoons 2Al(C_2H_5)_3 \xrightarrow{C_2H_4} 2AlR_3$$

$$- \frac{d[C_2H_4]}{dt} = k[Al(C_2H_5)_3]^{\frac{1}{2}}[C_2H_4]$$

There are three side reactions which occur during the growth step (8). At higher temperatures (>120°C), aluminum alkyls crack to form dialkylaluminum hydride and α-olefins (thermal displacement).

$$(RCH_2CH_2)_3Al \rightleftharpoons (RCH_2CH_2)_2AlH + RCH=CH_2$$

Because of the large excess of ethylene present in the growth reactor, the reverse reaction is insignificant. Ethylene reacts with dialkyl aluminum hydride much more rapidly than does the terminal olefin, and any alkyl group thermally displaced is replaced by an ethyl group. However, terminal olefin present in the growth reactor can react with trialkylaluminum compounds. The α-olefin inserts between the aluminum–carbon bond just as ethylene does in a normal growth process.

$$R_3Al + RCH=CH_2 \rightleftharpoons R_2AlCH_2-\underset{\underset{R}{|}}{CH}-R$$

This side reaction leads to the formation of branched alcohols as well as branched olefins. These impurities are dimeric and are about twice the average molecular weight as the product alcohols.

The third side reaction is formation of a small amount of polyethylene during the growth step. The quantity of polyethylene does not represent a significant yield loss, but does present serious processing problems. The polymer deposits on reactor surfaces, inhibits heat transfer, plugs valves, and must be cleaned out periodically. A Conoco patent (9) indicates this problem can be prevented by addition of small quantities of carbon monoxide to the feed ethylene.

The next step in a Ziegler process is oxidation of the trialkylaluminum growth mixture to the corresponding aluminum alkoxides.

$$2R_3Al + 3O_2 \longrightarrow 2(RO)_3Al$$

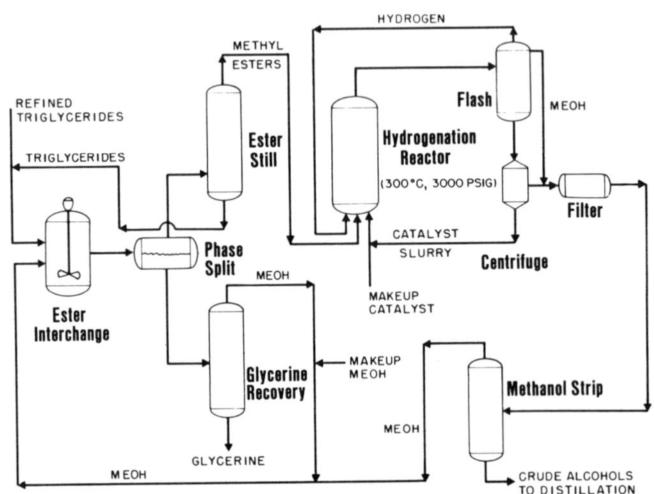

Figure 1.   Methyl ester hydrogenation

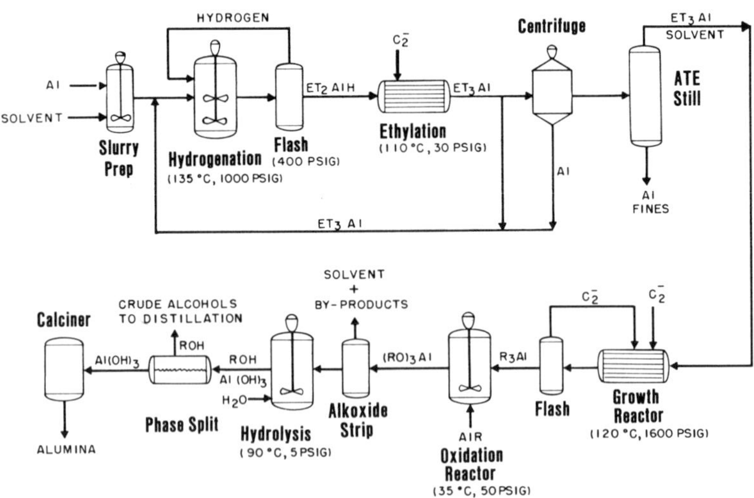

Figure 2.   ALFOL alcohol process

This step is carried out very carefully using dry air. The reaction is rapid and exothermic. By-products of the oxidation include aldehydes, esters, paraffins, and free alcohols.

The final step is hydrolysis of aluminum alkoxides to free the product alcohols.

$$(RO)_3Al + 3H_2O \longrightarrow 3ROH + Al(OH)_3$$

Hydrolysis can be carried out with water, as shown above, in which case the co-product is aluminum hydroxide. This aluminum hydroxide is converted to alumina by drying and calcining (10).

Aluminum alkoxides can also be hydrolyzed with dilute sulfuric acid in which case the co-product is aluminum sulfate.

$$2(RO)_3Al + 3H_2SO_4 \longrightarrow 6ROH + Al_2(SO_4)_3$$

Aluminum sulfate is a salable co-product in conventional markets, but it is of low value.

Conoco ALFOL Alcohol Process (11)

The ALFOL alcohol process is described in Figure 2. Conoco uses a two-step process for the synthesis of triethylaluminum--hydrogenation followed by ethylation. Triethylaluminum in a solvent and hydrogen are circulated through a hydrogenation reactor. Slurried aluminum powder is added and reacts with the triethylaluminum and hydrogen to form diethylaluminum hydride. The hydrogenation reaction is carried out at 135ºC and 1000 psig in an agitated vessel with an average residence time of one hour.

Powdered aluminum containing the patented (5) activators is available in railcar quantities. This powder is slurried with a dry solvent. Conoco uses a highly paraffinic solvent to ensure that product alcohols will meet FDA standards.

Effluent from the hydrogenation reactor is depressured to about 400 psig. This level of hydrogen is required to prevent the reverse reaction, diethylaluminum hydride decomposition, which results in plating of aluminum on the process equipment. Product diethylaluminum hydride, unreacted aluminum, and solvent are charged to the ethylation reactor. Ethylene is introduced and undergoes a rapid, exothermic reaction to form triethylaluminum. A tubular reactor with high heat transfer capabilities is required to control this reaction (12).

The triethylaluminum reaction product is divided into two streams. One stream, 70-75 percent of the total, is recycled directly to the hydrogenation unit to form additional diethylaluminum hydride. The other stream, 25-30 percent of the total, is the actual product stream. It is first centrifuged to remove the bulk of unreacted aluminum which is recycled to the hydrogenation reactor along with recycle triethylaluminum. Product triethylaluminum and solvent are then distilled from unreacted aluminum fines to produce high purity triethylaluminum.

The next step in the ALFOL alcohol process is chain growth. Triethylaluminum is preheated to 115°C in the first tubes of the growth reactor before ethylene is added. Ethylene is injected at several points along the length of the reactor to provide make-up ethylene as well as to help control the reaction temperature. Temperature control of this highly exothermic growth reaction is difficult and a special reactor design has been perfected to obtain good heat transfer.

The growth reaction is carried out below 130°C to prevent the alkyl decomposition or thermal displacement ($\underline{13}$) described earlier. Ethylene pressure is maintained at approximately 1600 psig. Temperature, pressure, and residence time are adjusted to obtain the desired extent of chain growth or "m-value." Excess ethylene is flashed from the trialkylaluminum product or "growth product" and recycled.

Oxidation of the trialkylaluminum mixture is carried out using dry air in agitated batch reactors in the ALFOL alcohol process. Reactor temperature is maintained at approximately 35°C by cooling and the oxidation is operated at about 50 psig. Approximately six hours is required for each batch of growth product to be completely oxidized. The oxidation reactors are operated in staggered fashion to minimize the maximum heat load.

The aluminum alkoxide mixture or "oxidized growth product" is fed to a series of vacuum flash evaporators to remove solvent introduced earlier in the triethylaluminum preparation. This vacuum stripping step also removes olefins formed during the growth reaction and the myriad of by-products formed during oxidation ($\underline{14}$). Efficiency of this stripping process is a key factor in alcohol product quality. This is the opportunity to separate volatile impurities--olefins, esters, aldehydes, paraffins, etc.--from product alcohols while the alcohols are in a nonvolatile form (aluminum alkoxides).

Stripped alkoxides are then sent to the hydrolysis reactor. In the current ALFOL alcohol process, hydrolysis is accomplished using water instead of dilute sulfuric acid which results in a mixture of alcohols and alumina slurry being formed in the hydrolysis reactor. This mixture is phase separated.

Alcohols are dried and sent to a distillation train where they are separated by conventional fractional distillation. Crude alcohols are separated into C_2-C_4, C_6-C_{10}, C_{12}-C_{14}, C_{16}-C_{18}, and C_{20}+ fractions. High purity, individual homologs are prepared by redistillation of the appropriate mixture. The product alcohols are marketed as ALFOL alcohols by Conoco Chemicals.

The alumina slurry is dried and then calcined to form a very active, high purity alumina which is marketed by Conoco Chemicals as CATAPAL alumina. Because of this unique process of manufacturing the alumina, it has a very low sodium content and almost no other trace metals are present. The alumina is in high demand as a catalyst support since it is nearly free of contaminants which can alter the performance of a catalyst. A catalyst manufacturer can attach only the metals he wishes to attach and be assured of their effects.

Ethyl Modified Linear Alcohol Process (15)

Ethyl's version of the Ziegler alcohol process has been modified in order to control the product alcohol distribution. Whereas the Conoco ALFOL alcohol process affords the full range of alcohols, C_2 –C_{30}, in a Poisson distribution, Ethyl's product distribution can be modified, for example, as shown in Figure 3 to give carbon number distributions to fit the needs of the market.

Table II

Typical Homolog Distributions

Carbon Homolog	Conoco ALFOL Process	Ethyl EPAL Process
2	0.5	Trace
4	3.4	0.1
6	9.5	1.5
8	16.1	3.5
10	19.5	8.0
12	18.4	34.0
14	14.1	26.0
16	9.1	16.0
18	5.1	8.8
20	2.5	1.9
22	1.1	0.2

The following discussion is based on an interpretation of patent literature. The Ethyl process consists of five basic steps instead of four.

1. Preparation of triethylaluminum.
2. Growth of ethylene to higher alkylaluminum compounds.
3. Transalkylation and separation of trialkylaluminum compounds.
4. Oxidation of aluminum alkoxides.
5. Hydrolysis to product alcohol.

Since the Conoco ALFOL alcohol process has already been described in detail, only those areas where the two processes are different will be covered. Preparation of triethylaluminum appears similar in both processes. Ethyl's scheme is believed to use a ballmilling procedure to obtain an active aluminum

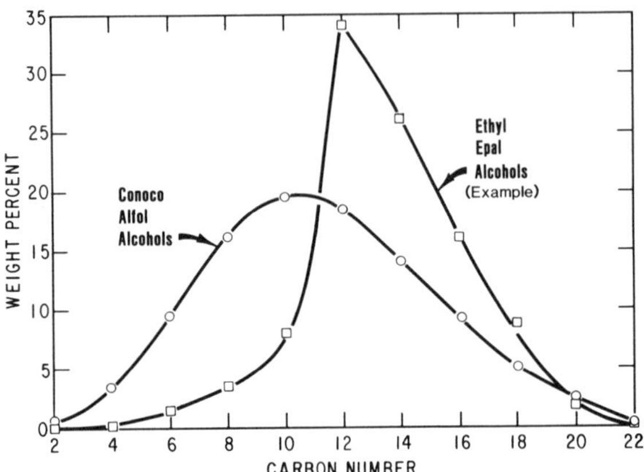

Figure 3.   Alcohol distribution

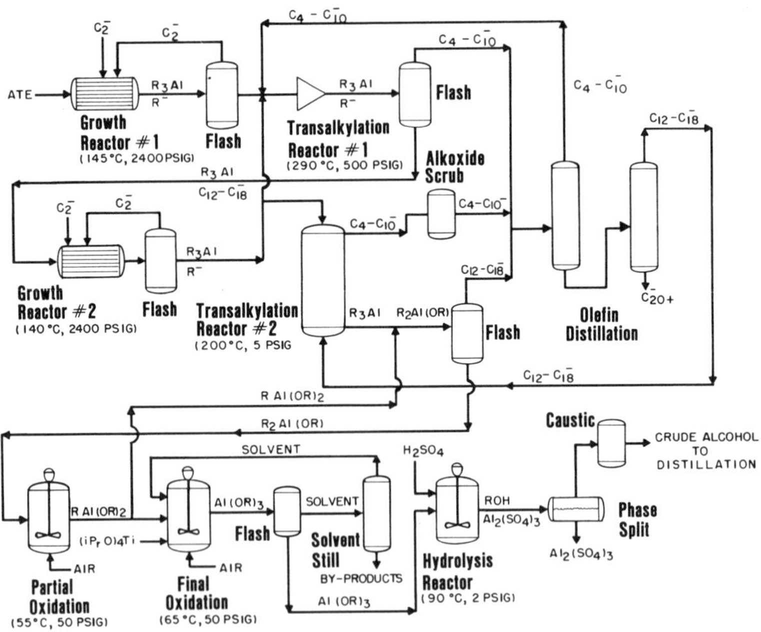

Figure 4.   Ethyl process (based on patent literature)

powder but the chemistry is identical. The Ethyl process, after triethylaluminum preparation, is shown in Figure 4. Ethyl patents (16) indicate two growth reactors, two transalkylation reactors, and two olefin stills are used to accomplish the desired alcohol peaking process.

In the first growth reactor, triethylaluminum is grown with ethylene as in the ALFOL alcohol process. However, the Ethyl growth process is extended to a lower "m-value" (3 vs. 4) and is carried out at higher temperature ($130°-150°C$) and pressure (2000–2500 psig). Exact reaction conditions will be determined by the "m-value" or alcohol distributin desired as well as the relative quantity of trialkylaluminum and olefins one wishes to produce (17). The higher growth temperature used in the Ethyl process will produce a significant quantity of lower molecular weight α–olefins via thermal displacement. These olefins are used to alter the product distribution of the aluminum alkyls and thus the alcohols. After growth, excess ethylene is removed in a flash drum and recycled.

The first growth product and its contained low molecular weight α–olefins are reacted with a large excess of α–olefins (C_4–C_{10}) in a transalkylation reactor. This reactor is a venturi-type reactor (18) operating at $275°-300°C$, 500 psig, and a short residence time of about 0.5 seconds. Heat is supplied to the reactor by super heating the large α–olefin stream. Immediate quenching of this product to about $120°C$ is required to suppress undesirable side reactions, such as dimerization and isomerization of the α–olefins.

The trialkylaluminum compounds leaving this transalkylation reactor now contain primarily C_4–C_{10} alkyl groups simply by the law of mass action. The low molecular weight olefins are separated from the trialkylaluminum compounds by distillation (19). This is a specially designed flash unit which contains provisions for scrubbing the overhead olefin vapors with partially (2/3) oxidized alkylaluminum dialkoxide to remove any trialkylaluminum carried overhead (20).

$$RAl(OR)_2 + R_3Al \longrightarrow 2R_2Al(OR)$$

Scrubbing provisions of this unit are not shown in Figure 4 for simplicity. Low molecular weight α–olefins are sent to olefin fractionation, and the low molecular weight trialkylaluminum compounds and high molecular weight olefins are sent to a second growth reactor.

Reaction conditions in the second growth reactor are similar to those in the first. Only a limited amount of chain growth occurs in this reactor such that the C_4–C_{10} trialkylaluminum compounds are grown to about C_8–C_{14} trialkylaluminum compounds. Again excess ethylene is removed in a flash drum and recycled.

This second growth product is fed to a second transalkylation reactor along with a $C_{12}-C_{18}$ α-olefin stream. Again, an Ethyl patent (21) indicates this is a countercurrent reactor in which the trialkylaluminum compounds are contacted with vapor phase α-olefins. The countercurrent design of this reactor permits the use of only a slight excess of detergent-range $C_{12}-C_{18}$ α-olefins instead of the large excess required by the cocurrent, venturi-type reactor. Reaction conditions are about 200 C and near atmospheric pressure (2-5 psig). Products exiting this reactor are trialkylaluminum compounds with primarily $C_{12}-C_{18}$ alkyl groups and a full range of α-olefins, C_4-C_{30}.

A slip-stream (1/5 to 1/3) of the second growth product can be sent to the first transalkylation reactor rather than the second. This procedure is required in order to balance this growth scheme.

Vapor-phase olefins from the second transalkylation reactor are scrubbed as before with alkylaluminum dialkoxide to remove any trialkylaluminum compounds and then sent to olefin distillation. Detergent-range trialkylaluminum compounds and high molecular weight olefins are blended with sufficient partially (2/3) oxidized growth product to convert the entire mixture to a blend having at least one alkoxide group per aluminum atom. This reaction is rapid and does not require a separate vessel, but can take place in the transfer lines. Conversion to monoalkoxide renders the aluminum compounds less volatile. High molecular weight olefins are then separated from the partially (1/3) oxidized growth product via a special flash unit similar to the one described earlier (20) and sent to olefin distillation.

The partially (1/3) oxidized growth product is sent to an oxidation unit where it is oxidized with dry air to an average of two alkoxide groups per aluminum atom (2/3 oxidized).

$$R_2Al(OR) + O_2 \longrightarrow RAl(OR)_2$$

This partial oxidation reactor is believed to operate at about 55°C. The partially (2/3) oxidized material is used as feed for the three scrubbers described above and for exchanging with unoxidized growth product to form a 1/3 oxidized product. The major use for the 2/3 oxidized growth product is as feed to the final oxidation reactor.

The final oxidation is a batch process as in the ALFOL alcohol process. A hydrocarbon solvent for viscosity control must be added prior to oxidation since none was introduced earlier. Based on patent literature (22), oxidation with dry air takes place in an agitated reactor at approximately 65° C and 50 psig in the presence of a titanium promoter which improves the oxidation selectivity.

The product aluminum trialkoxides are then stripped in a flash evaporator to remove solvent and oxidation by-products described earlier. The light hydrocarbon solvent is separated

from these by-products via distillation and recycled to the
oxidation reactor.

The aluminum trialkoxides are then hydrolyzed with dilute
sulfuric acid in the Ethyl process (23). This forms free alcohol and an aqueous aluminum sulfate solution which are separated by phase split. The aqueous aluminum sulfate is sold.
Product alcohols are washed with caustic to remove traces of
acid, dried, and fed to conventional distillation train. The
product alcohols are sold by Ethyl under the trade name of
EPAL alcohols.

The Ethyl EPAL process is more complex than the Conoco
ALFOL alcohol process. This complexity permits the flexibility of producing both α-olefins and alcohols from the same
processing unit as well as having considerable control over
the product homolog distributions. Penalties paid for this
flexibility are increased capital cost for a more complex process and production of some branched alcohols.

Literature Cited

1. Weast, R. C., editor, "Handbook of Chemistry and Physics," 51st Edition, The Chemical Rubber Company, Cleveland, OH, 1970, p. D-174.
2. Peters, R. A. "Kirk-Othmer Encyclopedia of Chemical Technology," 3rd Edition, Volume 1, John Wiley & Sons, NY, 1978, p. 716.
3. Monick, J. A. J. Amer. Oil Chemists Soc., 1979, $\underline{56}$ (11), 853A.
4. Ziegler, K. "Organometallic Chemistry," ACS Monograph No. 147, Van Nostrand, NY, 1960, p. 194.
5. Radd, F. J. and Woods, W. W. (to Conoco Inc.), U.S. 3,104,252 (1963).
6. Ziegler, K.; Gellert, H.; Zosel, K.; Holzkamp, E.; Schneider, J.; Soll, M.; and Kroll, W. Annalen, 1960, $\underline{629}$, 121.
6a. Weslau, H. Ibid, 198-206.
7. Allen, P. E.; Jones, G. R.; and Robb, J. C. Trans Faraday Soc., 1963, $\underline{63}$, 1936.
8. Gautreaux, M. F.; Davis, W. T.; and Travis, E. D. "Kirk-Othmer Encyclopedia of Chemical Technology," 3rd Edition, Volume 1, John Wiley & Sons, New York, 1978, p. 740.
9. Motz, K. L. and Lundeen, A. J. (to Conoco Inc.), U.S. 3,657,301 (1972).
10. Carter, W. B. (to Conoco Inc.), U.S. 3,264,063 (1966).
11. Lundeen, A. J. and Poe, R. L. "Encyclopedia of Chemical Processing and Design," 1974.
12. Lobo, P. A. (to Conoco Inc.), U.S. 2,971,969 (1961).
13. Lobo, P. A.; Coldiron, D. C.; Vernon, L. N.; and Ashton, A. T. Chemical Engineering Progress, 1962, $\underline{58}$ (5), 85.
14. Foster, V. and Acciarri, J. (to Conoco Inc.), U.S. 3,104,251 (1963).
15. Davis, W. T. (to Ethyl Corp.), U.S. 3,487,097 (1969).
16. Davis, W. T. (to Ethyl Corp.), U.S. 3,384,651 (1968).
17. Davis, W. T. and Kingrea, C. L. (to Ethyl Corp.), U.S. 3,415,861 (1968).
18. Davis, W. T. (to Ethyl Corp.), U.S. 3,345,476 (1969).
19. Gautreaux, M. F. (to Ethyl Corp.), U.S. 3,412,126 (1968).
20. Presswood, J. K. and Foster, W. E. (to Ethyl Corp.), U.S. 3,400,170 (1968).
21. Kottong, G. W. and Ritter, O. A. (to Ethyl Corp.), U.S. 3,389,161 (1968).
22. Cragg, H. J. and Nolen, D. A. (to Ethyl Corp.), U.S. 30,475,476 (1969).
23. Guzick, N. D. and McCarthy, J. H. (to Ethyl Corp.), U.S. 3,475,501 (1969).

RECEIVED March 2, 1981.

Applications of Higher Alcohols in Household Surfactants

TED P. MATSON

Conoco Inc., Chemicals Research Division, P.O. Box 1267, Ponca City, OK 74601

The major derivatives of normal primary higher alcohols used in the detergent industry include:

1. Alcohol ethoxylates (nonionics).
2. Alcohol ether sulfates.
3. Alkyl sulfates.
4. Miscellaneous--Including two other comparatively minor products--amine oxides and alkyl glyceryl ether sulfonates.

Smith([1]) gives an excellent review of commercial surfactants and others have reviewed the formulation of household products([2]).

Chemistry

Nonionics. The alcohol nonionics are processed according to the following:

$$ROH + n \ CH_2 \underset{\diagdown O \diagup}{-} CH_2 \longrightarrow RO(CH_2-CH_2O)_n H$$

The amount of ethylene oxide will generally be at least 60 percent of the total weight of the finished ethoxylate to give the best properties. This would be about 7 moles on a lauryl (C_{12}) range alcohol and 9 to 10 moles of ethylene oxide on a stearyl (C_{18}) range alcohol.

These resulting nonionics can be used in heavy-duty powders, heavy-duty liquids, and hard-surface cleaners.

Alcohol Ether Sulfates. The ether sulfates are prepared by the sulfation of alcohol ethoxylates as shown in the steps below:

$$RO(CH_2-CH_2O)_3H \xrightarrow[\text{2. NaOH}]{\text{1. } SO_3} RO(CH_2-CH_2O)_3SO_3Na + H_2O$$

The starting ethoxylate is generally the 3-mole adduct but can be as low as a 1-mole ethoxylate. Sulfating agents can be SO_3, chlorosulfonic acid, or sulfamic acid. A variety of neutralizing agents can be used--including NaOH, ammonia, triethanolamine, and $Mg(OH)_2$.

The ether sulfates are used in light-duty liquids, heavy-duty powders, and shampoos.

Alkyl Sulfates. The alkyl sulfates or alcohol sulfates are prepared very similarly to the ether sulfates shown above.

$$ROH \xrightarrow[\text{2. NaOH}]{\text{1. } SO_3} ROSO_3Na + H_2O$$

The alcohol sulfates are used in heavy-duty powders and shampoos.

Miscellaneous. The amine oxides used in household products are generally alkyl dimethyl amine oxides and are prepared from alcohols as shown below:

$$ROH \xrightarrow[\text{2. } (CH_3)NH]{\text{1. HCl}} RN(CH_3)_2$$

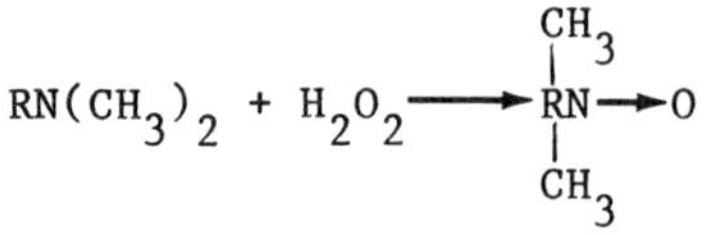

$$RN(CH_3)_2 + H_2O_2 \longrightarrow \overset{\displaystyle CH_3}{\underset{\displaystyle CH_3}{RN}} \rightarrow O$$

These alkyl dimethyl amine oxides are used as foam stabilizers primarily in light-duty liquids and shampoos where the major active ingredient is either an alcohol ether sulfate or an alcohol sulfate. The alkyl group is usually C_{12} to C_{14} which gives the best overall foam stability(3).

The alkyl glyceryl ether sulfonates are prepared as follows:

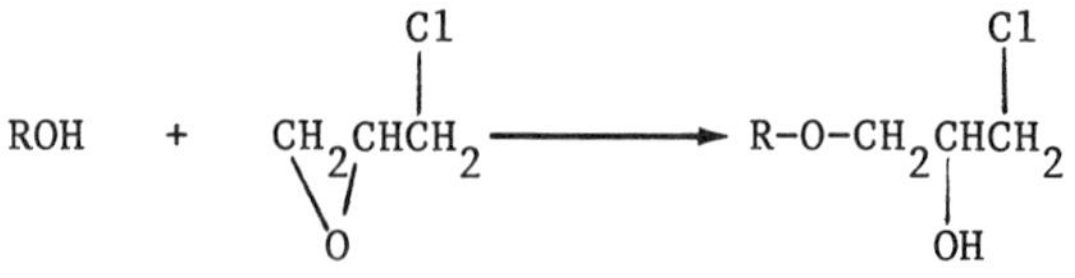

$$ROH + CH_2\underset{O}{\overset{Cl}{\underset{\diagdown\diagup}{CHCH_2}}} \longrightarrow R\text{-}O\text{-}CH_2\underset{OH}{\overset{Cl}{CHCH_2}}$$

Linear   Epichlorohydrin   Alkyl Ether
Alcohol   Chlorohydrin

$$R-O-CH_2\overset{\overset{\displaystyle Cl}{|}}{C}HCH_2 \;+\; Na_2SO_3 \longrightarrow R-O-CH_2\overset{\overset{\displaystyle SO_3Na}{|}}{C}HCH_2$$
$$\underset{\displaystyle OH}{} \qquad\qquad\qquad\qquad \underset{\displaystyle OH}{}$$

Alkyl Ether   Sodium   Alkyl Glyceryl
Chlorohydrin   Sulfite   Ether Sulfonate

These products can be found in light-duty liquids
and in some shampoos and detergent bars and are primarily
used to enhance skin emolliency and "cleaning ability."

Nonionics

Heavy-Duty Powders. Nonionics find use in the major
volume household product, heavy-duty powders. The words
heavy duty are synonymous with laundry applications. Products
of this type follow the general formulation as shown below:

10 percent nonionic.

25 to 35 percent sodium tripolyphosphate (or sodium
carbonate).

5 to 10 percent meta-silicate.

q.s. sodium sulfate and other minor constituent additives (bleach, perfume, antiredeposition agents, color,
etc.).

The optimum alcohol and amount of ethylene oxide is dependent upon the type of soil and the type of foam desired for
for the finished product. Figure 1($\underline{4}$) shows the optimum ethylene oxide content in a heavy-duty powder formulation similar
to that shown in the foregoing. Lines are "isodets"--lines
of equal detergency ranging from a lower detergency rating
of 1 to a high of 4.

Figure 1 shows that the optimum detergency is exhibited
in the 62 percent ethylene oxide range. On an alcohol
of C_{12} to C_{14}, this would amount to about 7 moles of ethylene
oxide. A higher alcohol molecular weight could result
in an ethylene oxide content approaching 10 moles. This
detergency was that exhibited on a combination of three
types of standard soiled cloths.

The optimum alcohol ethoxylate on oily soils can be
different. Nonionics have the advantage of performing
exceptionally well in comparison to other types of active
ingredients with oily soils. The choice of product can

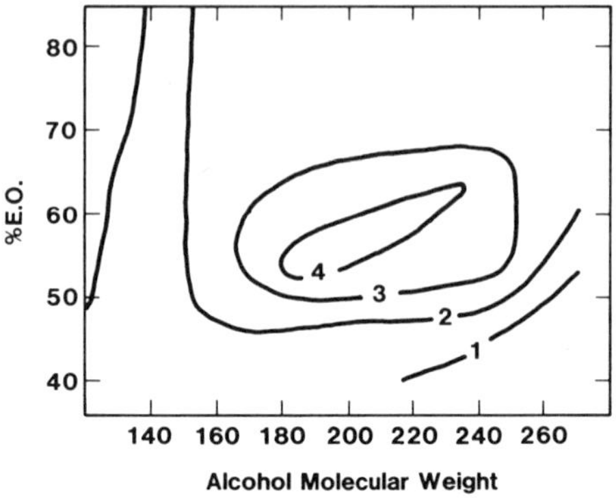

Figure 1.   Heavy-duty detergency alcohol, nonionic (0.2% concentration; 50 ppm hardness; 120°F)

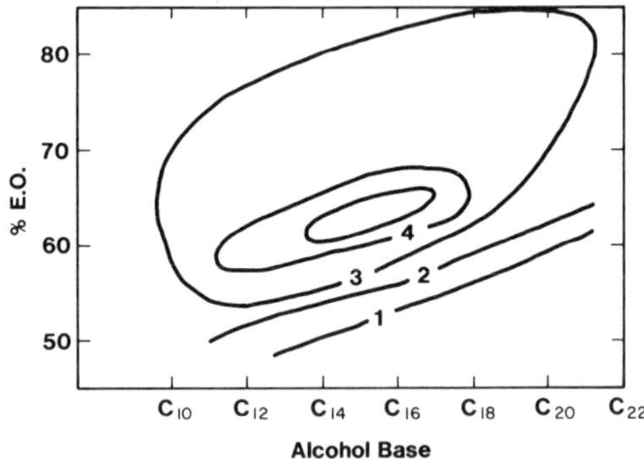

Figure 2.   Heavy-duty detergency nonionic isodets (0.4% concentration; 300 ppm hardness; 190°F)

also depend upon the temperature of use($\underline{5}$) as well as the
type of foam desired.

Figure 2 shows isodets at high temperature, high hardness, and high concentration giving a different optimum
than at the lower conditions of Figure 1. Foam and detergency profiles on different soil types with different wash
temperatures can be found in the literature($\underline{4}$-$\underline{10}$).

<u>Heavy-Duty Liquids</u>. Heavy-duty laundry liquids have
gained much importance and much interest in the literature
over the last few years. Products in the industry today
include both built and nonbuilt heavy-duty liquids. Built
heavy-duty liquids are those which contain a builder such
as tetrapotassium pyrophosphate (TKPP) or sodium citrate
in order to sequester calcium and magnesium hardness ions.
Nonbuilt heavy-duty liquids contain no builders and make
up for that shortage by considerably increased active contents in order to maintain performance($\underline{8}$-$\underline{14}$). Built heavy-
duty liquids can have either of the following types of
formulations:

10% Nonionic	10% Alkylbenzene sulfonate
10-20% TKPP or citrate	0-15% Nonionic
5-10% Hydrotrope	10-20% TKPP or citrate
q.s. H_2O	5-10% Hydrotrope
	q.s. H_2O

Nonbuilt heavy-duty liquids can contain either a high
ratio of nonionic to linear alkylbenzene sulfonate (LAS)
or, more recently, a reverse type ratio. Typical formulations of these types of products are given below:

30-35% Nonionic	or	25-35% LAS
10% LAS		10-15% Nonionic
5-10% Solubilizer		5-10% Solubilizer
45-55% H_2O		40-60% H_2O

Again, the optimum nonionic of choice for this application will depend upon the type of soil to be removed in
the laundry process. For example, Figure 3 shows the optimum
nonionic for removing typical sebum soil (body oil) in
a nonbuilt heavy duty liquid. This figure shows that the
optimum lies in the circle between C_{12} and C_{16} alcohol
at an ethylene oxide level of 60 to 80 percent. The peak
of this optimum would be in the vicinity of a C_{14} alcohol
with 70 percent EO. This is considerably higher in EO
content than the ethylene oxide optimum found for powdered
laundry detergents.

At the same time, as previously mentioned, different
types of soil can give different optimums. For example,

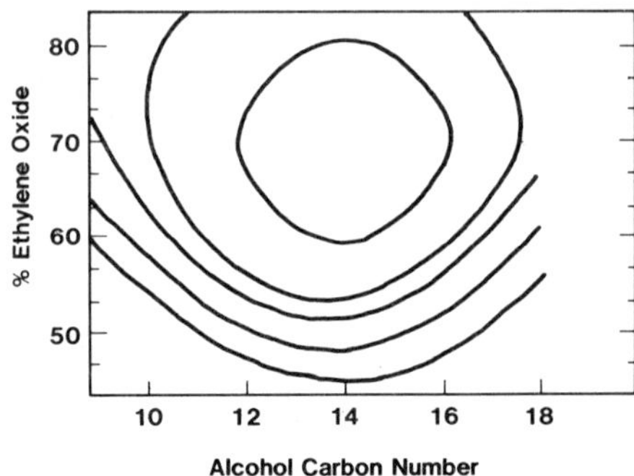

Figure 3.   Formulation 50% nonionic q.s. water (test conditions: 120°F; 150 ppm; 0.10% sebum soil)

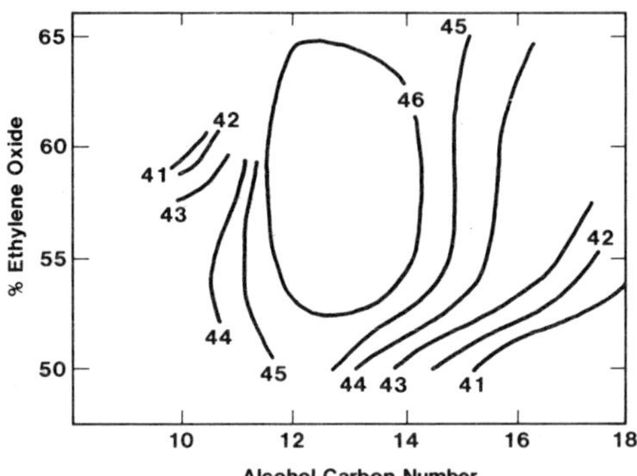

Figure 4.   Motor oil (300 ppm)

in Figure 4, we find that with motor oil soil, the optimum
detergency(46) is considerably narrower and is about a
C_{13} alcohol with 60 percent ethylene oxide.
      Most manufacturers are very interested in utilizing
optimum performances to make advertising claims and, therefore, will evaluate a large number of soils. The choice
of high nonionic/alkylbenzene sulfonate ratio as compared
to a reverse high alkylbenzene sulfonate/nonionic ratio
again depends upon the choice of foaming ability and the
choice of soils to be removed.
      The high nonionic formulation gives better performance
with sebum soils, whereas excellent performance on certain
oily soils can be obtained through the use of alkylbenzene
sulfonates at the high sulfonate to nonionic ratio.

      <u>General-Purpose Cleaners</u>. These are products which
can be used for a variety of household applications including
spray or bucket cleaners to be used to wash walls, woodwork,
porcelain, linoleum, etc. These products generally contain
much smaller amounts of active, as they are to be used
in a more concentrated form for eventual dilution with
water in actual application. These products can contain
5 to 10 percent nonionic in the total liquid and may also
contain TKPP as well as certain solvents such as ethylene
glycol mono n-butyl ether. The general choice of products
for this application would be the standard C_{12} to C_{13} alcohols
with 60 to 70 percent ethylene oxide. However, a C_{10} alcohol
ethoxylate offers excellent grease cutting and hard surface
cleaning properties.

Ether Sulfates

      <u>Heavy-Duty Powders</u>. Although the number of products
in which one finds alcohol ether sulfates in laundry products
is exceptionally small, the volume is fairly high due to
the volume of those specific products. The ether sulfate
which is generally used in laundry powders is based on
a C_{13} to C_{15} average alcohol with 1 to 3 moles of ethylene
oxide on the average. Figure 5 shows this to be the highest
detergency. Due to the need for spray-drying, these will
always be the <u>sodium</u> alcohol ether sulfate. The ether
sulfate is normally found in these formulations at a 4
to 5 percent level and is used in conjunction with alkyl-
benzene sulfonate and/or alcohol sulfates in the formulation.
The specific use in this type of product was discussed
at the AOCS World Conference(2).

      <u>Light-Duty Liquids</u>. Light-duty liquids refer to the
liquid products sold today primarily for the end application
of dishwashing. They can also be used for the washing

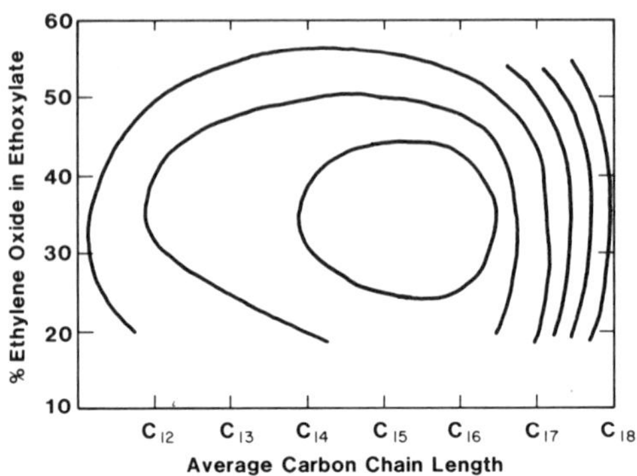

Figure 5.   Ether sulfate detergency (0.12%; 150 ppm; 110°F)

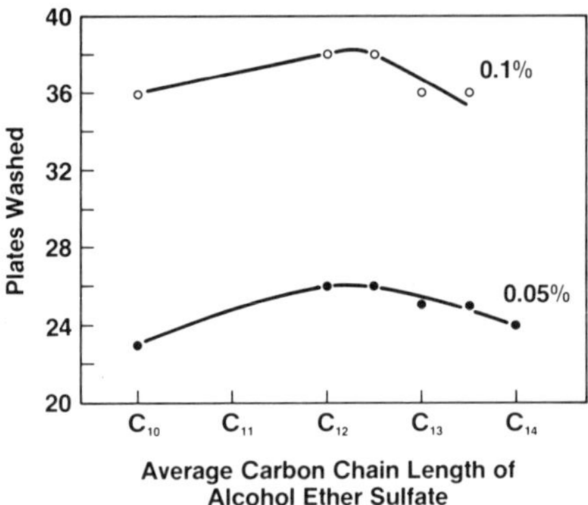

*Figure 6.   Optimum ether sulfate for ABS/ES light-duty liquids (24/6 LAS/
AES—50 ppm hardness)*

of fine fabrics. Light-duty liquids generally fall into
two major categories. One of these includes the alcohol
ether sulfate as the total major active ingredient. In
the other type of formulation, one finds alkylbenzene sulfonate (LAS) and ether sulfate. LAS and ether sulfates work
in conjunction with one another to develop a synergistic
effect upon foam stability and, as a result, are widely
found in the commercial liquid formulations today. The
typical ether sulfate product used in either of these types
of liquids is a lauryl-range alcohol with about a 3-mole
ethylene oxide content (some products may have a mixture
of alcohol sulfate and ether sulfate, but in general the
overall average EO content is still in the 2- to 3-mole
range). Figure 6 shows the optimum carbon chain length
for an ether sulfate and substantiates the use today of
the C_{12} to C_{13} range. Going above that level, one does
not have too much drop off in performance, but this would
result in considerably less solubility of the total formulation. Going below the C_{12} optimum will cause a drop in foam
stability performance(15).

The following table shows the effect of ethylene oxide
upon the solubility of the formulation. From these results,
it can be readily understood why the 40 percent ethoxylate
(3 mole) is usually used for this application.

EFFECT OF EO CONTENT ON SOLUBILITY OF LD LAS/ES LIQUIDS

ES Portion of Active	Percent EO	Cloud Point ($^{\circ}$F)
C_{12}-C_{14}	17.1	48
C_{12}-C_{14}	33.5	34
C_{12}-C_{14}	39.8	30

The optimum ratio of LAS to ether sulfate for foam
stability alone is generally in the range of 4:1 LAS/ether
sulfate. This can be seen in Figure 7 where the average
number of plates washed is plotted against different LAS/ES
ratios.

Shampoos. Ether sulfates are extensively used in
shampoo formulations in the range of 15 to 25 percent active.
The general product of choice is the lauryl-range alcohol
with 1 to 3 moles of ethylene oxide. In shampoos, either
sodium, ammonium, or triethanolamine salts of the ether
sulfate can be used. The choice of these products will
be primarily related to the solubility and the finished
form of the shampoo. Triethanolamine salts are considerably
more soluble and, thus, will be found in a large number

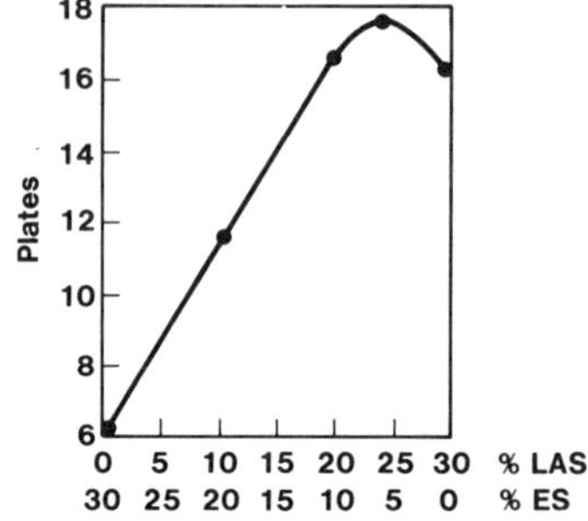

Figure 7. Foam stability—LAS/ES (0.05% concentration; 50 ppm hardness; 115°F, 46°C)

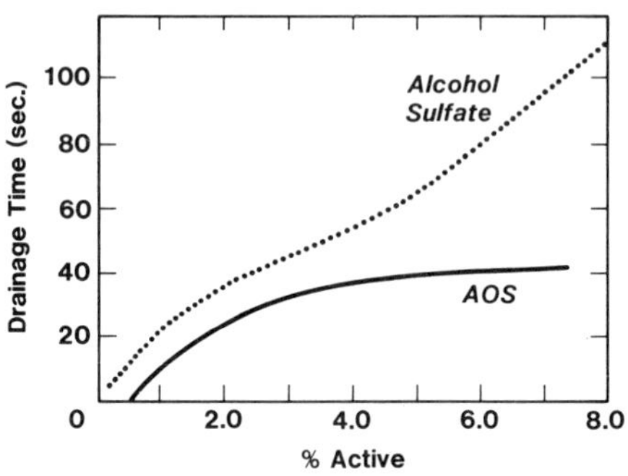

Figure 8.   Drainage time at 50 ppm hardness sodium alcohol sulfate vs. AOS

of liquid products (especially clear liquids). As one
moves towards a lotion, cream, or even paste shampoo, products move towards the sodium salt. These shampoos will
generally include foam stabilizers to further enhance the
foam capacity. These foam stabilizers are normally alka-
nolamides or amine oxides. The use of the ether sulfate
along with these foam stabilizers gives thick copious foam,
and this foam is the general property desired for shampoos.
There has been over the last few years considerable use
of ether sulfates made from 1 to 1 1/2 moles of ethylene
oxide on the alcohol. The philosophy behind this use is
that one can obtain the thicker, more copious foam of an
alcohol sulfate with lesser EO on the alcohol and yet still
maintain considerable solubility (more like the ether sulfate).

Alcohol Sulfates

A major application for alcohol sulfates is its use
in conjunction with other active ingredients in heavy-duty
laundry powders. In these products, as mentioned before
under ether sulfates, one will usually find LAS as well
as ether sulfates. Ten to fifteen years ago, a considerable
amount of alcohol sulfate was used in conjunction with
LAS. These products had a very high content of sodium
tripolyphosphate which sequestered the hardness and, as
a result, allowed the alcohol sulfate to do its exceptional
job of detergency(16). As phosphate contents have been
limited, the use of the alcohol sulfate has been reduced
due to its sensitivity to harder waters. Another factor
which has somewhat slowed down the use of alcohol sulfates
is the gradual lowering of wash temperatures from an average
of $140^{\circ}F$ many years ago (and even boiling temperatures
before that time) down to an average of $100^{\circ}F$ or less today.
The alcohol sulfate of choice today ranges between
a carbon chain average of C_{13} to C_{17}. This alcohol sulfate
can be found in some laundry products at 3 to 5 percent
of the total formulation.

Shampoos. Like ether sulfates, alcohol sulfates used
in shampoos are generally in the range of C_{12} to C_{14} (the
lauryl alcohol range). The alcohol sulfate gives an exceptionally thick foam and gives that foam over a wide range
of soil concentrations. The alcohol sulfates are the choice
product over olefin sulfonates due to the exceptional foam-
ability at the actual higher use concentrations found in
washing the hair at high-soil loads (Figure 8). Figure 8
shows this foam stability advantage as a function of film
drainage at different shampoo concentrations.

Miscellaneous

As mentioned earlier, the amine oxides are used in dishwashing liquids and in shampoos where the major surfactant is either an alcohol ether sulfate or an alcohol sulfate.

Alkyl glyceryl ether sulfonates are found in some dishwashing liquids, shampoos, and detergent bars and are primarily used to enhance skin emolliency though some claims are made for "cleaning ability."

Abstract

The major derivatives of normal primary higher alcohols used in the detergent industry were discussed. These included alcohol ethoxylates, alcohol ether sulfates, and alkyl sulfates. The chemical reactions for preparation of each were also given.

The applications of each of these derivatives were discussed with typical formulations and optimum carbon chain lengths presented. Usage of alcohol derivatives in laundry products, dishwashing liquids, and shampoos were highlighted.

Literature Cited

1. Smith, G. D. "Solution Chemistry of Surfactants," 1, K. L. Mittal, editor. Plenum Publishing Corp., 1979.
2. Matson, T. P. J. Am. Oil Chemists Soc., 1978, 55, 66.
3. Matson, T. P. J. Am. Oil Chemists Soc., 1963, 40, 640.
4. Matson, T. P. Soap Chem. Specialties, November 1963, 39, 52.
5. Matson, T. P. Specialities, 1966, 2, 17.
6. Schick, M. J., Editor "Nonionic Surfactants," Marcel Dekker, Inc., New York, NY, 1966.
7. Schonfeldt, N., "Surface Active Ethylene Oxide Adducts," Pergamon Press, Oxford and London, 1969.
8. McKenzie, D. A. J. Am. Oil Chemists Soc., 1978, 55, 93.
9. McGuire S. E.; Matson, T. P. J. Am. Oil Chemists Soc., 1975, 52, 11.
10. Dillan, K. W.; Goodard, E. D.; McKenzie, D. A. J. Am. Oil Chemists Soc., 1979, 56, 59.
11. Matson, T. P.; Berretz, M. Soap Chem. Specialities, November 1979, 55, 33.
12. Matson, T. P.; Berretz, M. Soap Chem. Specialties, December 1979, 55, 41.
13. Matson, T. P.; Berretz, M. Soap Chem. Specialties, January 1980, 56, 36.
14. Matson, T. P.; Berretz, M. Soap Chem. Specialties, February 1980, 56, 41.
15. Stanberry, C. J.; Matson, T. P.; Langford, J. J. Soap Chem. Specialties, 1964, 40, 43.
16. Matson, T.P. J. Am. Oil Chemists Soc., 1963, 40, 636.

RECEIVED December 29, 1980.

Secondary Alcohol Ethoxylates

Physical Properties and Applications

NAOJI KURATA[1], KAZUO KOSHIDA, HIROMI YOKOYAMA, and TAKAKIYO GOTO

Technical Department of Kawasaki Plant, Nippon Shokubai Kagaku Kogyo Co., Ltd., 10-12 Chidori-cho, Kawasaki-shi, Japan

Only a few companies in the world have produced surfactant range higher secondary alcohols until now. The first secondary alcohol plant appeared in the U.S.S.R. in 1959 and has been in operation since then. It has a capacity of about 10,000 tons per year of alcohols having carbon numbers ranging from 13 to 17. The alcohols are mainly used to produce sulfosuccinate esters through esterification with maleic anhydride followed by addition of sodium sulfite. The sulfosuccinate esters are used as raw materials for powder detergents there. Meanwhile, in 1964 in the U.S.A., Union Carbide Corporation commercialized a series of ethoxylated nonionic surfactants based on their secondary alcohols having carbon numbers ranging from 11 to 15 under the registered trade name of TERGITOL 15-S. The plant is believed to have a capacity of 36,000 tons per year of 3 mole ethoxylate of the alcohols. The third plant for higher secondary alcohols was constructed by Nippon Shokubai in Japan in 1972. The plant capacity, originally 12,000 tons per year of 3 mole ethoxylate, was expanded to 18,000 tons per year in 1977 and now is being further expanded to 30,000 tons per year. Nippon Shokubai's products are also sold mainly in the form of ethoxylate under the registered trade name of SOFTANOL. The alkyl carbon range of SOFTANOL is from 12 to 14 so far.

All industries, especially the surfactant and detergent industries, have been heavily involved in various recent controversies such as health and safety problems, resource and energy conservation movements and so on. Under these circumstances, more emphasis has to be placed on key words: SAFETY, RESOURCE SAVING and EFFICIENCY. Only by satisfying these three will there be the chance of developing new technologies and new products.

It is hoped that this presentation will give some ideas for more beneficial use of the products derived from higher secondary alcohols to those who are seeking new technical development in the surfactant and detergent industries.

[1] Current Address: Nippon Shokubai K.K. Co., Ltd., Osaka-shi, Japan.

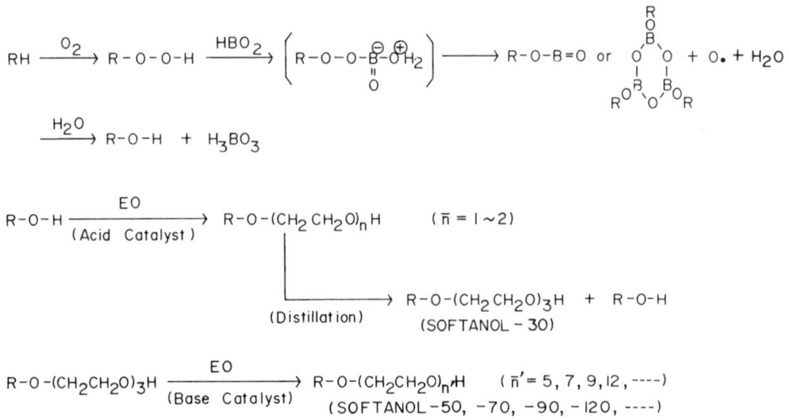

$$RH \xrightarrow{O_2} R-O-O-H \xrightarrow{HBO_2} \left(R-O-O-\overset{\ominus}{\underset{O}{B}}-\overset{\oplus}{O}H_2 \right) \longrightarrow R-O-B=O \text{ or } \cdots + O_\bullet + H_2O$$

$$\xrightarrow{H_2O} R-O-H + H_3BO_3$$

$$R-O-H \xrightarrow[\text{(Acid Catalyst)}]{EO} R-O-(CH_2CH_2O)_nH \quad (\bar{n} = 1{\sim}2)$$

$$\xrightarrow{\text{(Distillation)}} R-O-(CH_2CH_2O)_3H + R-O-H \quad \text{(SOFTANOL - 30)}$$

$$R-O-(CH_2CH_2O)_3H \xrightarrow[\text{(Base Catalyst)}]{EO} R-O-(CH_2CH_2O)_{n'}H \quad (\bar{n}' = 5, 7, 9, 12, \text{----})$$
$$\text{(SOFTANOL -50, -70, -90, -120, ----)}$$

Figure 1. *Synthesis route to secondary alcohol ethoxylates from* n-*paraffins*

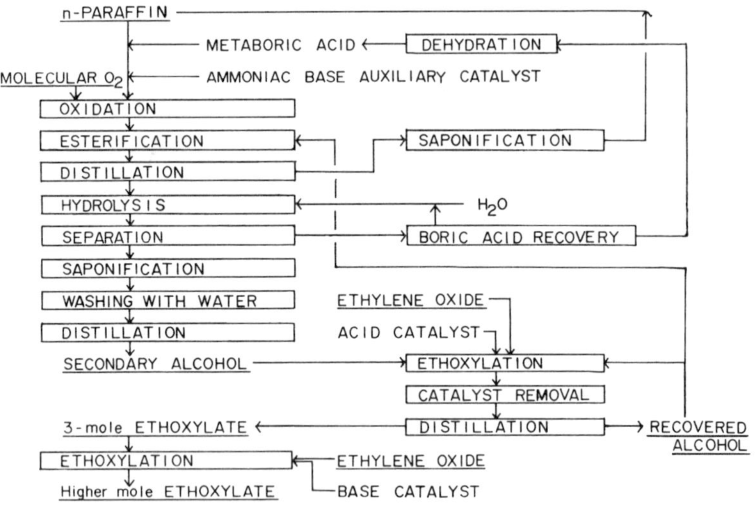

Figure 2. *Simplified Nippon Shokubai's process route for secondary alcohols and their ethoxylates*

Manufacturing Process of Secondary Alcohols and Their Ethoxylates

Since previous papers([1],[2]) describe details of the manufacturing process for secondary alcohols(SA) and their ethoxylates
(SAE), only the outline of the process will be presented here. A
mixture of secondary alcohols is obtained by liquid phase air-
oxidation of normal paraffins in the presence of a boric acid
catalyst(Figure 1). Although the existing commercial processes,
as developed independently, comprise significantly different com-
binations of various unit processes, they are all based on this
boric acid-modified oxidation of hydrocarbons([3]).

In Nippon Shokubai's process(Figure 2), the 3 mole ethoxylate
of a mixture of secondary alcohols can be produced from a mixture
of normal paraffins through a fully integrated continuous process.
The oxidation is carried out using β-metaboric acid as a catalyst
and an ammoniac base as an auxiliary catalyst to promote the
reaction([4]). The alcohol mixture obtained consists of all possible structural isomers of secondary alcohols having the same
carbon numbers as the raw material used.

The ethoxylation of secondary alcohols must be first carried
out using an acid catalyst to a low degree of polymerization ([2]).
The product, 3 mole ethoxylate, is separated from the reaction
mixture by stripping the unreacted alcohol. Further ethoxylation
can then be carried out just as with the primary alcohol or alkylphenol using a base catalyst in the conventional manner.

In this process for 3 mole ethoxylate, a mixture of n-paraffins having three successive carbon numbers can be used. The
spread of four or more successive carbon numbers may cause difficulties in the separation and purification steps.

Table I shows up-to-date process economics for C_{12-14} secondary alcohols and their 3 mole ethoxylate.

Considerations of Carbon Number Ranges

n-Paraffins are usually isolated from kerosene and have carbon numbers ranging from 10 to 16. As a raw material for Nippon
Shokubai's SOFTANOL, C_{12-14} was chosen from the viewpoints of feed
stock availability and versatile performance of the products.

Table II shows a typical example of carbon number distribution of n-paraffins in kerosene now available in Japan. To utilize all n-paraffins in kerosene in the near future, it is convenient to divide them, for instance, into three fractions each having three successive carbon numbers, i.e. C_{10-12}(ave. 11), C_{12-14}
(ave. 13), and C_{14-16}(ave. 15).

Because of great differences in their vapor pressures(Table
III), differences in the process variables originating from carbon
number distribution of raw materials are relatively great in this
process, especially in the oxidation step. A comparison is summarized in Table IV, which shows the disadvantage of a lower carbon
number range to some extent.

Table I   Process Economics for C_{12-14} Secondary Alcohols
          and for Their 3-mole Ethoxylate

<u>Raw Material and Utilities Consumption, (per ton of product)</u>

	Alcohol	3-mole Ethoxylate
Normal Paraffins	1,240 kg	750 kg
Ethylene Oxide	----	420 kg
Other Chemicals	10,000 yen	8,000 yen
Steam	3.3 ton	3.0 ton
Fuel	----	----
Electric Power	500 kwh	400 kwh
Process Water	3.3 m^3	2.5 m^3
Cooling Water	600 m^3	500 m^3
Inert Gas	10 m^3	10 m^3

<u>Commercial Installation</u>
12,000 tons/yr of C_{12-14} Secondary Alcohols and 18,000
tons/yr of 3-mole Ethoxylate at Nippon Shokubai's Kawasaki
Plant, Japan.

Table II   Typical Example of Carbon Number Distribution
of n-Paraffins Extracted from Kerosene

Carbon Number	C_{10}	C_{11}	C_{12}	C_{13}	C_{14}	C_{15}	C_{16}
Distribution(%)	7	25	24	22	15	6	1

Table III   Vapor Pressure and Boiling Point of n-Paraffins

Carbon Number	C_{11}	C_{13}	C_{15}
Vapor Pressure at 170°C(torr)	380	130	44
Boiling Point(°C)	196	235	271

Typical Physical and Surface Active Properties

To examine the physical and surface active properties of SAE,
three secondary alcohol samples each having three successive
carbon numbers, as mentioned earier, were prepared. For practical
reasons, blends of alcohols are chosen instead of alcohols having
individual carbon numbers. The alcohol samples were ethoxylated
to various degrees of polymerization for testing. Two previous
papers by other workers are recommended with reference to this
subject. One by MacFarland(5) of Union Carbide Corp. deals with
a blend of C_{11-15} secondary alcohols and the other by Matson(6)
of Continental Oil Co. deals with individual carbon chain
homologs.

Pour point, viscosity, cloud point, wetting power and foam
properties, being important advantages of SAE, are presented
here in comparison with other commercial products derived from
primary alcohols (Ziegler and Oxo) or nonylphenol (branched
chain).

Pour Point. Figure 3 shows pour points vs. Griffins' HLB
value for various ethoxylated nonionics.

SAE, in general, have far lower pour points than those of
primary alcohols(PA). Among three secondary alcohol series with
different carbon number ranges, the lower the carbon number range,
the lower the pour point, especially in the lower HLB or water-
insoluble region.

Nonylphenol ethoxylates(NPE) have pour points similar to
those of SAE in the higher HLB region but differ in the HLB region
below 10.

Smith(7) and Fisher(8) recently published articles on the relationship of viscosity, pour point and alkyl chain length of
primary alcohol ethoxylates(PAE) and SAE. The empirical equations
proposed are in fairly good agreement with the authors' obser-

Table IV   Comparison of Process Variables with Different Carbon Number Ranges for the Production of Sec.-Alcohols

Factors	Carbon Number Range of n-Paraffins		
(Oxidation Step)	$C_{10} - C_{12}$	$C_{12} - C_{14}$	$C_{14} - C_{16}$
Reaction Temperature	Normal	Normal	Normal
Reaction Pressure	Higher	Normal	Lower
Conversion	Lower	Normal	Higher
Selectivity	Lower	Normal	Higher
Space Time Yield	Lower	Normal	Higher
(Separation and Purification Step)			
Distillation Temperature	Lower	Normal	Higher
Distillation Pressure	Lower	Normal	Higher
Manufacturing Cost	Higher	Normal	Lower

Table V   Viscosity and Gel Range of Aqueous Surfactant Solutions

cP at 25°C

Surfactant	Surfactant Concentration (% by weight)									
	10	20	30	40	50	60	70	80	90	100
C_{10-12}SAE (7EO)	2	7	28	75	98	94	84	77	65	50
" (9EO)	2	5	18	68	151	142	112	97	81	64
" (12EO)	2	6	28	118	221	221	171	134	114	94
C_{12-14}SAE (7EO)	7	27	77	132	160	3×10^3	1×10^4	105	81	52
" (9EO)	4	8	40	230	1×10^3	270	7×10^3	130	96	65
" (12EO)	5	9	37	270	G E L	G E L	300	180	130	87
C_{14-16}SAE (7EO)	3	23	74	135	162	6×10^3	1×10^4	100	77	52
" (9EO)	2	6	35	218	2×10^5	258	1×10^4	123	93	61
" (12EO)	2	6	30	229	G E L	G E L	227	176	126	87
C_{12+14}PAE (7EO)	4	28	715	G E L	G E L	G E L	5×10^4	1×10^5	104	56
" (9EO)	3	6	37	G E L	G E L	G E L	2×10^3	192	123	69
" (12EO)	3	8	53	G E L	G E L	G E L	G E L	310	158	SOLID
N P E (8EO)	167	292	421	597	8×10^3	7×10^4	1×10^5	2×10^5	274	232
" (10EO)	4	32	320	G E L	G E L	2×10^3	9×10^4	404	290	257
" (13EO)	2	6	40	G E L	G E L	G E L	882	408	306	279

vations on SAE within the range of C_{10} to C_{16} and EO mole numbers of below 15. Their equations can be used to predict pour point or viscosity of SAE from their chemical structures.

Viscosity. Figure 4 shows plots of viscosity at 25°C vs. EO mole numbers for the three series of SAE. Here, the lower the carbon number range, the lower the viscosity.

PAE have much higher viscosities than SAE having the same carbon number range. Thus, the viscosity of C_{12+14} Ziegler PAE is about the same as that of C_{14-16} SAE except in the higher EO mole region, where PAE (12EO) is solid at this temperature. NPE have still higher viscosities than SAE.

As to the viscosity of aqueous solutions of nonionic surfactants, in general, gel formation tendency is very important. Table V and Figure 5 show differences in viscosity of aqueous solutions and gel ranges for various nonionics. Gel ranges of SAE are remarkably narrow compared with those of PAE and NPE(Table V).

Cloud Point. Figure 6 shows plots of cloud point against Griffins's HLB for various nonionics. As the cloud points of the three series of SAE are closely on a line within this HLB region, the cloud point of a SAE can be estimated from a curve made with other SAE having different alkyl carbon number ranges. PAE have higher cloud points than SAE at the same HLB. This might mean that secondary alkyls behave as stronger hydrophobes than do primary alkyls. NPE have different slopes of cloud point vs. HLB curves compared with alcohol ethoxylates.

Wetting Power. Figure 7 shows plots of wetting time against Griffin's HLB for various nonionics. Better wetting abilities of SAE compared with PAE are seen as in previous papers by MacFarland(5) and by Zika(9, 18).

Figure 8 shows temperature dependence of the wetting power of PAE, SAE and NPE each having about the same calculated HLB. It indicates a lower wetting power of PAE at lower temperatures and of NPE at higher temperatures compared with that of SAE.

Foam Properties. A previous paper(1) reported that SAE show better foam breakability than PAE or NPE. This tendency becomes clearer when SAE and PAE having lower carbon number ranges are compared. Figure 9 and Figure 10 show foam volume vs. concentration in an agitation test for C_{12-14} SAE (9EO), C_{10-12} SAE(9EO) anc C_{9+11} PAE(8EO) in the initial stage and after five minutes, respectively. Figure 11 shows the time required for 25% foam reduction vs. concentration of surfactant. In these figures, we can see that C_{10-12} SAE shows rapid foam reduction at lower or higher concentration. This should be a noteworthy phenomenon among various types of nonionic surfactants and can be applied in the formulation of easy rinsing detergents. A reason for such a maximum in the curves of foam volume vs. concentration has not yet been found.

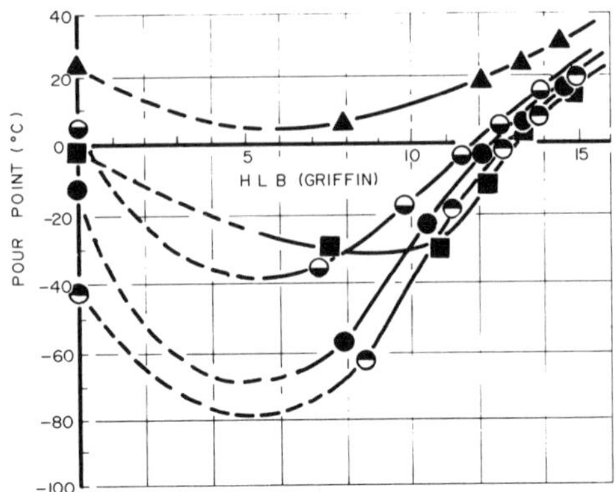

Figure 3.   Pour point vs. HLB for various ethoxylated nonionics ((–◐–) C_{10-12} SAE; (–●–) C_{12-14} SAE; (–◑–) C_{14-16} SAE; (–▲–) C_{12+14} PAE (Ziegler); (–■–) NPE)

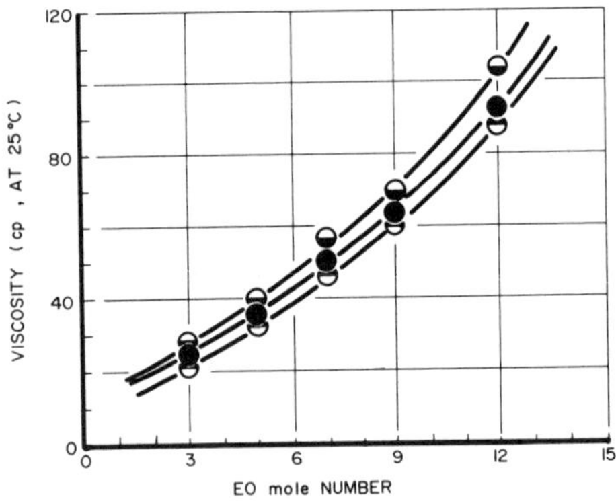

Figure 4.   Viscosity vs. EO mole number for secondary alcohol ethoxylates ((–◐–) C_{10-12} SAE; (–●–) C_{12-14} SAE; (–◑–) C_{14-16} SAE)

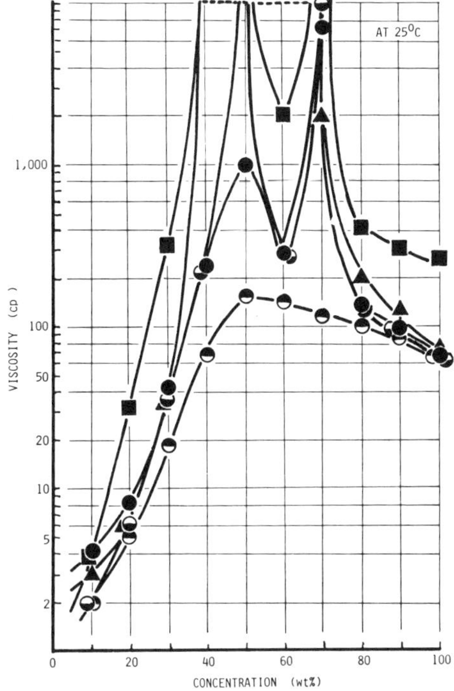

Figure 5. Viscosity vs. concentration for various ethoxylated nonionics ((−◒−) C_{14-16} SAE (9EO); (−●−) C_{12-14} SAE (9EO); (−◓−) C_{10-12} SAE (9EO); (−▲−) C_{12+14} PAE (9EO) (Ziegler); (−■−) NPE (10EO))

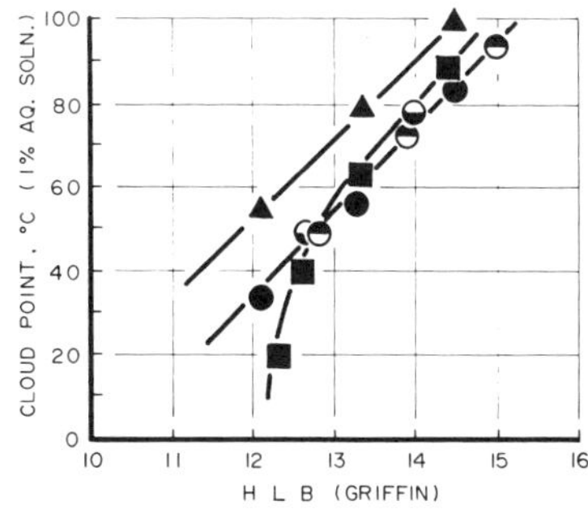

Figure 6. Cloud point vs. HLB for various nonionics ((−◓−) C_{10-12} SAE; (−●−) C_{12-14} SAE; (−◒−) C_{14-16} SAE; (−▲−) C_{12+14} PAE (Ziegler); (−■−) NPE)

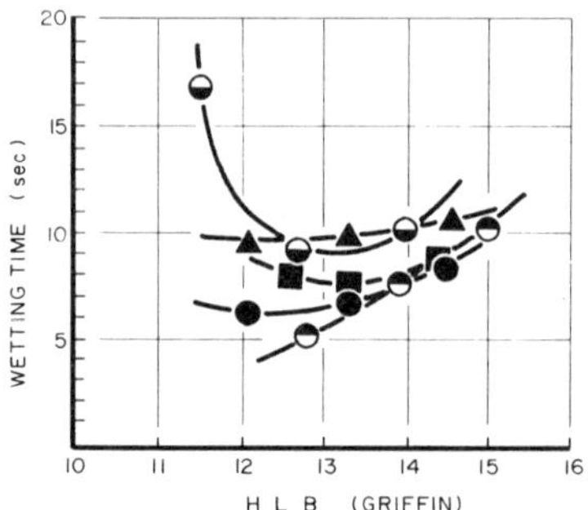

Figure 7. Wetting time vs. HLB for various ethoxylated nonionics. Test conditions: method—JIS-K-3362-1965; roller-cloth wool—10 × 90 mm; surfactant concentration—0.1 wt %; temperature—25°C. ((–◒–) C_{10-12} SAE; (–●–) C_{12-14} SAE; (–◓–) C_{14-16} SAE; (–▲–) C_{12+14} PAE (Ziegler); (–■–) NPE)

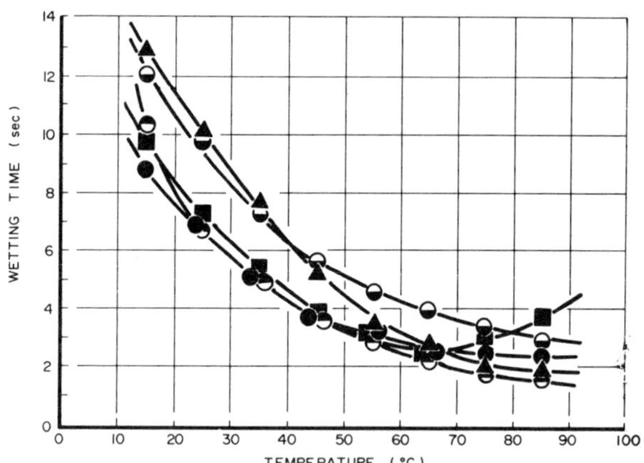

Figure 8. Wetting time vs. temperature for several nonionics. Test conditions: method—JIS-K-3362-1965; roller-cloth wool—10 × 90 mm; surfactant concentration—0.1 wt %. ((–◒–) C_{10-12} SAE (8EO), HLB = 13.4, CP = 58°C; (–●–) C_{12-14} SAE (9EO), HLB = 13.3, CP = 56°C; (–◓–) C_{14-16} SAE (10EO), HLB = 13.2, CP = 54°C; (–▲–) C_{12+14} PAE (9EO), HLB = 13.3, CP = 79°C (Ziegler); (–■–) NPE (10EO), HLB = 13.3, CP = 63°C)

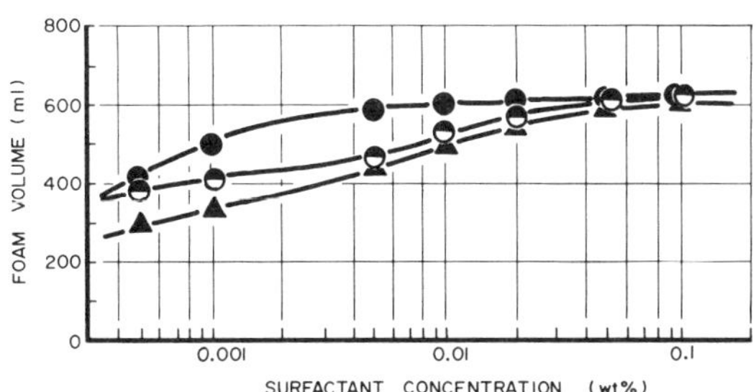

Figure 9. Effect of carbon number range on foam volume (initial stage) ((–◖–) C_{10-12} SAE (9EO); (–●–) C_{12-14} SAE (9EO); (–▲–) C_{9+11} PAE (8EO))

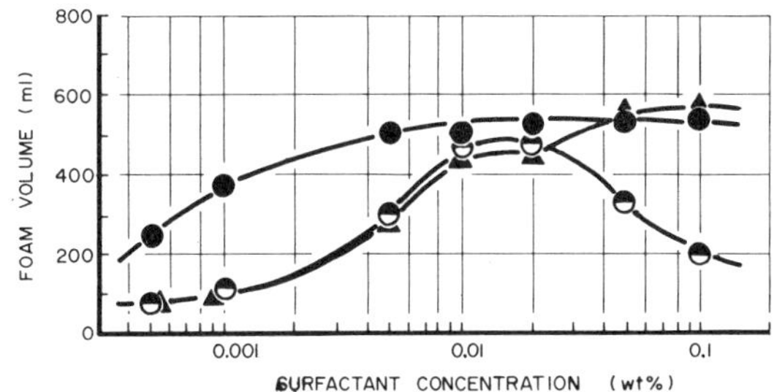

Figure 10. Effect of carbon number range on foam volume (after 5 min) ((–◖–) C_{10-12} SAE (9EO); (–●–) C_{12-14} SAE (9EO); (–▲–) C_{9+11} PAE (8EO))

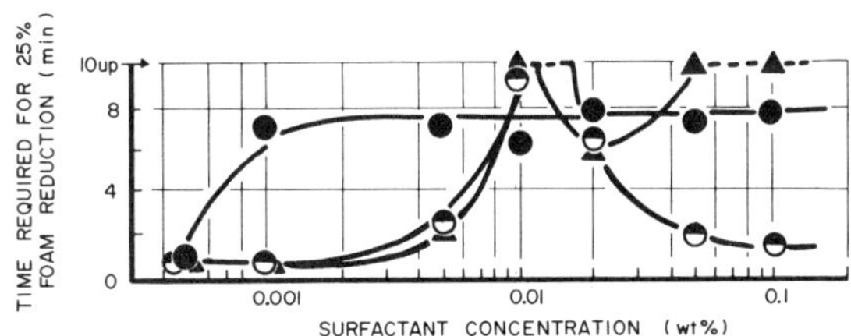

Figure 11. Effect of carbon number range on foam reduction rate ((–◖–) C_{10-12} SAE (9EO); (–●–) C_{12-14} SAE (9EO); (–▲–) C_{9+11} PAE (8EO))

Good foam breakability of SAE having a lower carbon range
may lead to excellent performance in detergent formulations which
other surfactants have never shown.

Safety and Environmental Assessment

Biodegradability. The biodegradability of SAE has already
been extensively examined by other workers using various test
methods such as the standard BOD test, Shake-Culture test,
Warburg test, River Die-away test, and Continuous Activated-
Sludge test. All of these laboratory tests(10) showed rapid
degradation of SAE. Several results obtained in the authors'
laboratory have been reported, a summary of which follows.

a. River Die-away test(11). To examine biodegradation in
natural rivers, water samples were taken from the upper, the
middle and the lower regions of the Tama River, flowing along the
border between two big cities, Tokyo and Kawasaki(Figure 12).
The biodegradabilities were followed by three analytical
methods, i.e. (A) foam height method, (B) cobalt thiocyanate
colorimetry (CTAS) and (C) COD (Cr) method.
Table VI shows the analytical data for the river water imme-
diately after sampling. Although the sampling points were located within a distance of only about 30 km, the river water clearly
showed different degrees of pollution. It seemed, therefore,
very interesting to see how different bacterial populations and
nutritive concentrations might affect the speed of biodegradation.
Three typical commercial samples tested are shown in Table
VII. Figure 13, Figure 14 and Figure 15 show the test results
obtained using the river water taken from Noborito, Fussa and
Ohme, respectively.

Table VI   River Water Used in the Test *(11)*

Sampling Point	Sampling Date	COD(Cr), ppm	CTAS, ppm	Foam, ppm
(1) Noborito	Apr. 21, '73	12	0	0
(2) Fussa	May 31, '73	4	0	0
(3) Ohme	Aug. 8, '73	0	0	0

Yukagaku

Table VII   Nonionic Surfactant Samples Used in the River
             Die-away Test *(11)*

NPE(10EO)	Branched-chain nonylphenol 10 mole ethoxylate
PAE(9EO)	C_{12} - C_{15} Oxo alcohols (consisting of 60% straight- and 40% branched-chain) 9 mole ethoxylate
SAE(9EO)	C_{12} - C_{14} Secondary alcohol (consisting of all possible position isomers) 9 mole ethoxylate

Yukagaku

Figure 12. Map of the Tama River basin (11)

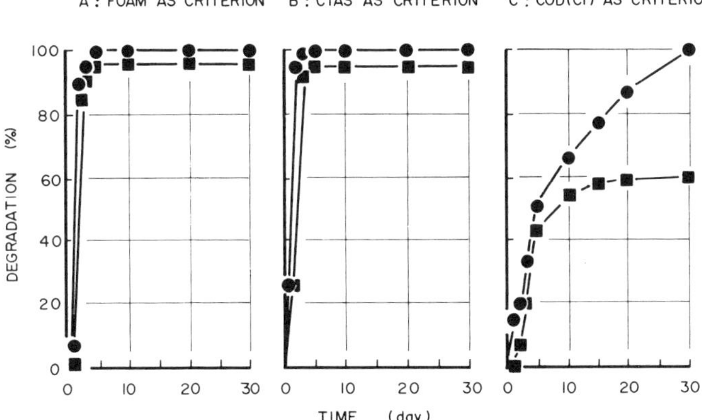

Figure 13. Biodegradability of nonionic surfactants in river die-away test using the Tama River water, Noborito ((–●–) C_{12-14} SAE (9EO); (–■–) NPE (10EO))
(11)

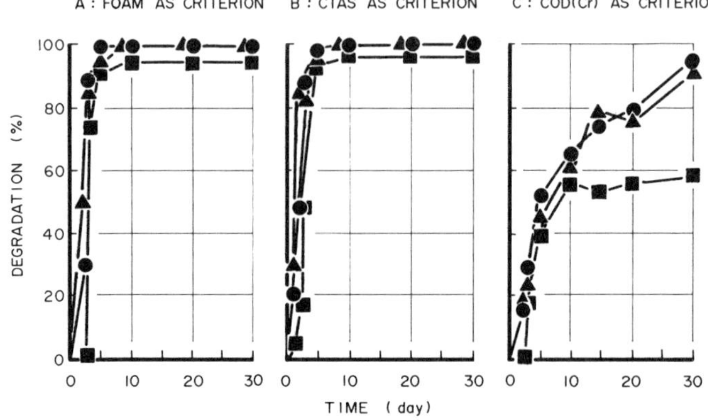

Figure 14. Biodegradability of nonionic surfactants in river die-away test using the Tama River water, Fussa ((–●–) C_{12-14} SAE (9EO); (–▲–) C_{12-15} PAE (9EO), (Oxo); (–■–) NPE (10EO)) (11)

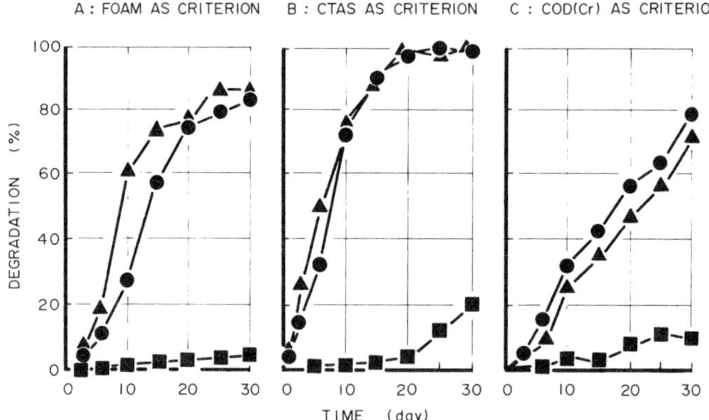

Figure 15. Biodegradability of nonionic surfactants in river die-away test using the Tama River water, Ohme ((–●–) C_{12-14} SAE (9EO); (–▲–) C_{12-15} PAE (9EO), (Oxo); (–■–) NPE (10EO)) (11)

As seen in Figure 13, in river water from the lower regions, both SAE(9EO) and NPE(10EO) seemed to be degraded in a relatively short period, with no difference in foam height and with little difference by CTAS. However COD(Cr) measurements showed that about 40% of the NPE(10EO) remained undegraded after 10 days, and even over a prolonged period.

As seen in Figure 14, which shows data taken using river water from the middle regions, the rate of degradation was a little slower than in the lower regions in all three analytical measurements, with a tendency about the same as in the case of the lower regions. With regard to foam height and CTAS, PAE(9EO) and SAE(9EO) showed almost the same behavior. In the case of COD (Cr) measurements, the alcohol-based surfactants showed the same degradation rate while decomposition of NPE(10EO) did not exceed about 60%, as in the lower regions.

In Figure 15, which shows data from the upper regions, the Ohme, it is seen that, even in the case of foam height and CTAS, NPE(10EO) was resistant to decomposition. Thus, it was found that the NPE was extremely slow to be decomposed compared with alcohol-based nonionics, in the clean water of the upper regions. Moreover, even in the highly polluted water of the lower regions, it seemed to be decomposed only with regard to surface activity but never reached ultimate decomposition. On the contrary, alcohol-based nonionics, derived from both primary and secondary alcohols, seemed to be completely degraded in all cases..

The foregoing is in line with earlier work which showed that acclimated river water degraded an alkyl phenol ethoxylate when measured by foam volume reduction(11a).

b. Continuous-flow activated sludge test(12). This test was carried out to show that SAE is by far more easily degradable than NPE in a continuous-flow activated sludge system.

The apparatus used is the same as that adopted by Conway et al(13). The test was conducted over long periods at higher concentration of surfactants than in any previous studies. In the test, biodegradation was followed by measuring COD(Cr), COD(Mn), cobalt thiocyanate active substances (CTAS) and TOC(Figure 16). It was found that SAE(9EO), contrary to NPE(10EO), was adequately biodegraded by any measurements used, in spite of the high concentration of surfactant (200mg/1).

Toxicity to Fish(14). As to the relationship between surface activity and fish toxicity of surfactants, it is said that at a surface tension of water below 48 dynes/cm fish cannot breathe through their gills, resulting in fatal toxicity due to the changes in physical properties.

We examined the relationship between changes in surface tension and changes in toxicity and studied the toxicity of biodegradation intermediates of surfactants.

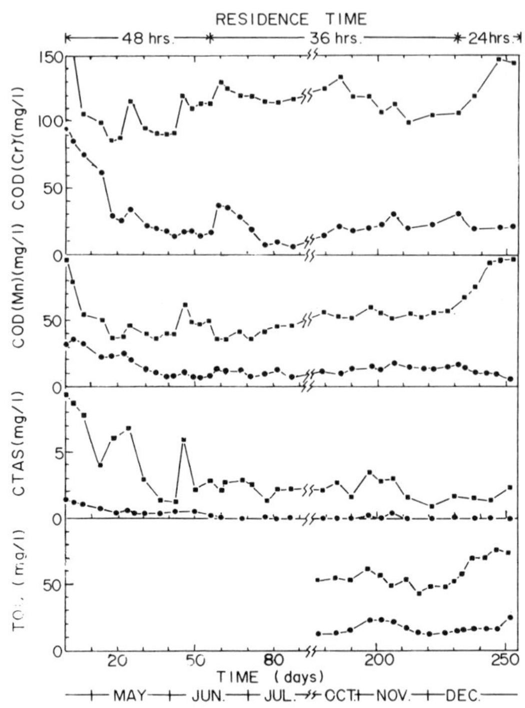

Test Conditions

Sample		NPE(10EO)	SAE(9EO)
Influent Composition			
Surfactant	(mg/l)	200	200
Urea	(mg/l)	25	25
Phosphoric Acid	(mg/l)	10	10
Ave. Anal. Value in Influent			
COD(Cr)	(mg/l)	339	406
COD(Mn)	(mg/l)	210	156
CTAS	(mg/l)	200	200
TOC	(mg/l)	120	123
Residence Time	(hrs)	48–24	48–24
MLSS	(mg/l)	2000±500	2000±500

Yukagaku

Figure 16. COD(Cr), COD(Mn), CTAS, and TOC remaining in continuous-flow activated sludge effluents fed 200 mg/L of NPE(10EO) or SAE(9EO) ((–●–) C_{12-14} SAE (9EO); (–■–) NPE (10EO)) (12)

a. Acute toxicity (TLm48). Table VIII shows the test results for acute toxicity of various surfactants on gold fish.

The results show the same tendency as previously in that the stronger the surface activity, the stronger the toxicity, and the higher the ethylene oxide mole ratio, the lower the toxicity. TLm values obtained for secondary alcohol-based surfactants are also found to obey the known general rules. Macek and Krzeminski(14a) also reported a similar conclusion on acute effects of APE, PAE and SAE to bluegill sunfish by their static bioassays.

It is apparent that the toxicity of LAS is much stronger than that of ethoxysulfates derived both from primary and secondary alcohols.

Table VIII Acute Toxicity to Fish *(14)*

Surfactant	TLm(48hr), mg/l[*]
Nonioncs	
NPE(10EO)	5.4
C_{12-14} SAE(7EO)	3.3
C_{12-14} SAE(9EO)	5.1
C_{12-14} SAE(12EO)	12.0
C_{12} PAE(9EO)	1.9
C_{12-15} PAE(9EO)(60% straight/40% branched chain)	1.4
Anionics	
LAS	4.7
sec.-AES(Sulfate of C_{12-14} SAE(3EO))	43.0
prim.-AES(Sulfate of Lauryl Alc.-(2EO))	36.0

[*]Surfactant concn. to kill 50% of fish within 48 hr.
Method : Based on JIS-K-0102   Fish : 5 cm Goldfish

b. Toxicity after biodegradation. Table IX and Table X are the results of preliminary tests of aquatic toxicity of various surfactants during their biodegradation in river water. Very high initial concentrations of surfactants were adopted here in order to clarify the differences in degree of toxicity reduction.

Table IX shows that 5 days aging of river water containing C_{12-14} SAE(9EO) or C_{12} PAE(9EO) can remarkably reduce their toxicity to fish even at the initial concentration of 20 mg/l, which corresponds to several times their TLm, and thus indicates that their toxicity tends to be reduced with their biodegradation

Table IX   Toxicity to Fish after 5d Biodegradation for Nonionic Surfactant (14)

Surfactant	Initial Concentration mg/l	% Fish Alive		
		After 2h	After 24h	After 48h
SAE(9EO), C_{12-14}	20	100	90	70
PAE(9EO), C_{12}	20	100	100	60
NPE(10EO)	20	0	0	0
Control	0	100	100	90
NPE(10EO)	5	40	20	0

Yukagaku

Table X   Toxicity to Fish After 5 and 10d Biodegradation for Anionic Surfactants (14)

Surfactant	Initial Concentration mg/l	% Fish Alive					
		5d Biodegradation			10d Biodegradation		
		after 2h	after 24h	after 48h	after 2h	after 24h	after 48h
SAE(3EO)S[a]	100	100	90	90	100	100	100
	200	40	0	0	100	100	100
LAE(2EO)S	100	100	70	40	100	80	80
	200	0	0	0	0	0	0
LAS	100	0	0	0	0	0	0
	200	0	0	0	0	0	0
Control	0	100	100	90	100	100	100

a   Sulfate of C_{12-14} Secondary Alcohol 3EO
b   Sulfate of Lauryl Alcohol 2EO

Yukagaku

processes. On the contrary, NPE(10EO) shows no reduction in
toxicity even at the initial concentration of 5 mg/l, suggesting
that the toxicity of its biodegradation intermediates should be
marked.

Table X shows similar results for anionic surfactants, in
which no reduction in toxicity of LAS was observed but some reduction was found in the case of C_{12-14} SAE(3EO) sulfate and C_{12}
PAE(2EO) sulfate especially at their initial concentration of
100 mg/l.

c. Surface tension and toxicity of biodegradation intermediates. Table XI shows the relationship between the ratio of live
fish, surface tension and CTAS remaining after biodegradation.
It indicates that both SAE(9EO) and PAE(9EO) have no toxicity at
3 - 4 mg/l CTAS remaining but NPE(10EO) still has strong
toxicity at 2.8 mg/l CTAS remaining. Thus, it was found that the
TLm_{48} values of both PAE and SAE after 4 days biodegradation in
river water were 91 mg/l based on the initial concentration, and
that of NPE was 3.7 mg/l (Table XII). In the case of NPE, the
observed value, 3.7 mg/l, was lower than its original TLm_{48}.
This means that the toxicity of its biodegradation intermediates
should be higher than that of the original surfactant.

Figure 17 shows the surface tension after 4 days biodegradation vs. % live fish after 48 hrs. In the case of PAE and SAE,
a surface tension of 48 dynes/cm after 4 days biodegradation was
found high enough for fish to live, while in the case of NPE, 51
dynes/cm was found too low for fish to live. The results on NPE
here are somewhat different from those of Macek and
Krzeminski(14a), who reported that bluegill sunfish are probably
less susceptible to the acute effects of APE than of AE in their
dynamic bioassay conditions. They also reported in determining
toxicity of the biodegradation products of four surfactants,
including NPE(9EO), no mortality was observed even at an initial
surfactant concentration ranging from 32 to 40 ppm. These
differences might be explained by differences in biodegradation
routes under different conditions and test procedures, or from
differences in the alkyl structure of the NPE used.

Performance in Household Detergents

SAE now have a variety of applications in both household and
industrial fields, some typical examples of which follow. Before
going into details, some advantageous characteristics of SAE,
which should be taken into account by users, are summarized in
Table XIII.

Table XI   Toxicity of Biodegradation Intermediates of
           Nonionic Surfactants to Fish (14)

Nonionic Surfactant	Intial Concentration mg/1	Intial Surface Tension (dyne/cm)	CTAS After 4d Biodegradation (mg/1)	Surface Tension After 4d Biodegradation (dyne/cm)	% Fish Alive		
					After 24hr	After 48hr	After 72hr
C_{12-14} SAE(9EO)	35	33.9	0.4	58	100	100	100
	50	32.5	1.2	52	100	100	100
	75	31.5	3.7	48	100	100	100
	110	30.9	9.3	41	0	0	0
	160	30.9	18.2	38	0	0	0
C_{12-15} PAE(9EO) Oxo	35	34.2	0.3	60	100	100	100
	50	33.5	1.0	55	100	100	100
	75	33.0	2.9	48	100	100	90
	110	32.8	6.5	42	10	0	0
	160	32.7	15.0	38	0	0	0
NPE(10EO)	2	53.2	1.0	63	100	100	100
	3	51.1	1.8	57	100	100	80
	4.5	49.0	2.8	51	20	0	0
	6.5	47.0	3.9	47	0	0	0
	9.5	45.1	6.0	40	0	0	0

Table XII   TLm (48hr) of Biodegradation Intermediates
            Calculated from Initial Surfactant Concentration (14)

Surfactant	Original TLm(48hr) (mg/1)	TLm (48hr) After 4 Day Biodegradation* (mg/1)
SAE(9EO)***	5.1	91
PAE(9EO)**	1.4	91
NPE(10EO)	5.4	3.7

*   Initial surfactant concentration to kill 50% of the fish
     within 48 hr.
** C_{12-15} Oxo-alcohol 9 mole ethoxylate, consisting of 60%
     straight - and 40% branched - chain
*** C_{12-14} Secondary alcohol 9 mole ethoxylate, consisting of
     all possible position isomers

Table XIII   Characteristics of Secondary Alcohol Ethoxylates

Properties	Advantageous Effects
Low Pour Point	Easy Handling
Low Viscosity	
High Solubility	Easy Liquid Formulation
Narrow Gel Range	
Good Wetting Power	High Performance
Good Foam Breakability	
Good Biodegradability	Easy Waste Treatment
Containing no Free Alcohol	Odorless

Detergency – HD Laundry Application. To evaluate the detergency of three series of SAE, each having a different carbon no. range, typical Terg-O-Tometer tests were run on artificially soiled cotton and polyester/cotton. Test conditions and soil type are summarized in Table XIV. The results are shown in Figure 18 and Figure 19. As seen in these figures, the higher the carbon number range of SAE, the higher the detergency. This tendency is interesting if applied to low or no phosphate detergents. But it should be noted that the detergency obtained here represents removal of clay-based soil. In removal of more oily soil, the tendency may be somewhat different.

Figure 20 shows detergency vs. EO mole unit for C_{12-14} SAE with or without builder. It is seen that the builder has considerable efficiency in removal of clay-based soil.

Figure 21 shows a comparison of detergency of SAE(9EO) and LAS at different STPP concentrations. This result is also interesting in the formulation of non-built liquid detergents.

More detailed studies on detergency of SAE of single carbon no. chain ranging from C_{10} to C_{16} were made by Matson[6]. From his studies, it can be said that an optimum choice of SAE in non-built HDL should be in the range C_{12-14} with 60 – 64% EO (7 – 9EO). An interesting test result on the removal of mineral oil from a polyester substrate was reported by Dillan et al (15). This suggests that SAE has much higher detergency for polyester fabric than PAE at lower temperatures.

Detergency in Dish Washing. Detergency tests for dish washing were conducted by the Modified Leenerts test prodecure. The test conditions and apparatus are shown in Table XV.

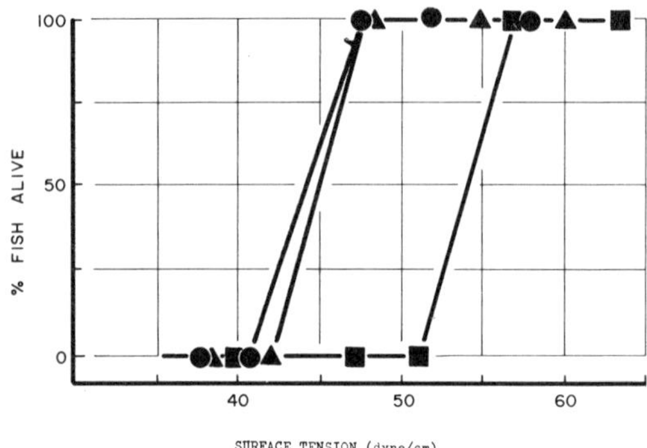

*Figure 17. The % fish alive (48) vs. surface tension after 4 day biodegradation ((-●-) C_{12-14} SAE (9EO); (-▲-) C_{12-15} PAE (9EO), (Oxo); (-■-) NPE (10EO))
(14)*

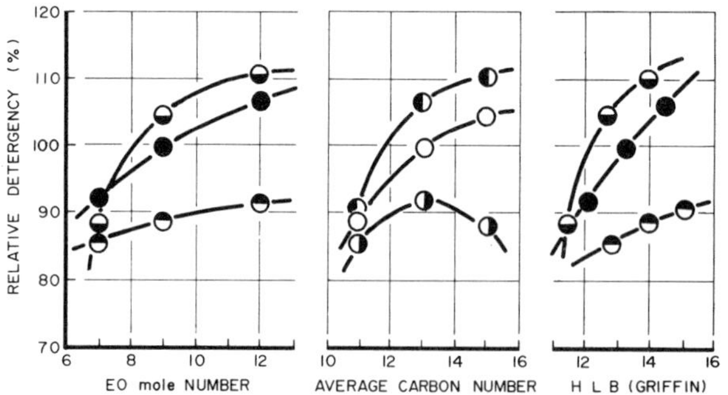

Figure 18. Relationship between carbon number range and detergency for cotton broadcloth. Test condition: surfactant concentration—0.03 wt %. ((-◐-) C_{10-12} SAE; (-●-) C_{12-14} SAE; (-◒-) C_{14-16} SAE; (-◑-) SAE (7EO); (-○-) SAE (9EO); (-◐-) SAE (12EO))

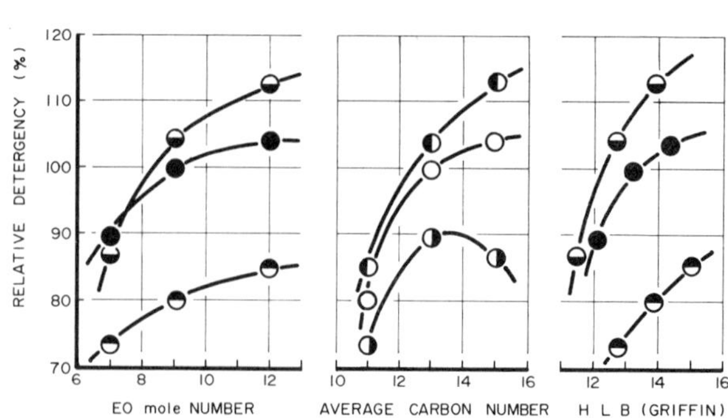

Figure 19. Relationship between carbon number range and detergency for poly-ester/cotton broadcloth. Test condition: surfactant concentration—0.03 wt %. ((-◐-) C_{10-12} SAE; (-●-) C_{12-14} SAE; (-◒-) C_{14-16} SAE; (-◑-) SAE (7EO); (-○-) SAE (9EO); (-◐-) SAE (12EO))

Table XIV Terg-O-Tometer Test Conditions

Temperature	$25 \pm 2°C$
Agitation Speed	100 rpm
Washing Time	5 min.
Rinsing Time	5 min.
Water Hardness	50 ppm
Liquid Volume/Pot	1 Liter
Size of Cloth Swatch	10 X 10 cm
Fabric Load/Pot	3 Swatches
Reflectance of Soiled Swatch	Rs = ca. 50%
Points of Rs Detection	3

Soil Composition ; (A) : (B) : (C) = 0.5 : 49.75 : 49.75
  (A) Carbon : Standard of the Japan Oil Chemists' Society
  (B) Inorganic Component : The Kanto District Loam,
                    250 Mesh Passed
  (C) Organic Component :

Organic Soil Components,	wt%
Myristic Acid	16.7
Oleic Aicd	16.7
Tristearin	16.7
Triolein	16.7
Cholesterol	8.8
Cholesteryl Stearate	2.2
Paraffin Wax(mp:50-52°C)	11.1
Squalene	11.1

Soil Penetration ; Modified Supersonic Wave Method Using
        CCl_4 as Solvent

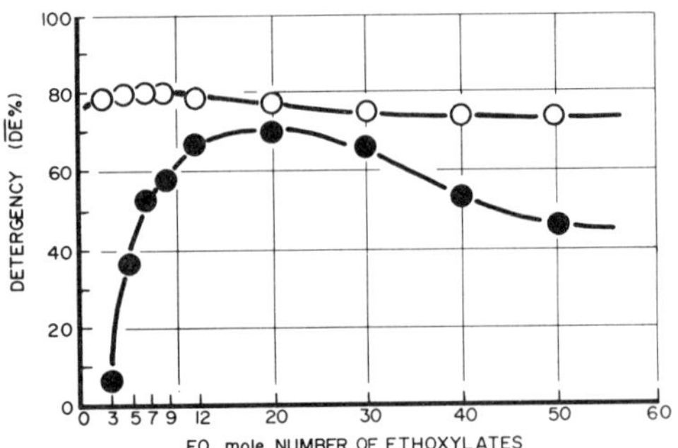

Figure 20.   Detergency of C_{12-14} secondary alcohol ethoxylates with or without builders. Test conditions: surfactant concentration—0.03 wt %; builder concentration—0.12 wt %; builder composition: STPP—20 wt part; Na_2CO_3—5 wt part; sodium metasilicate—10 wt part; Na_2SO_4—20 wt part; CMC—1 wt part. ((–●–) C_{12-14} SAE alone; (–○–) C_{12-14} SAE and builder)

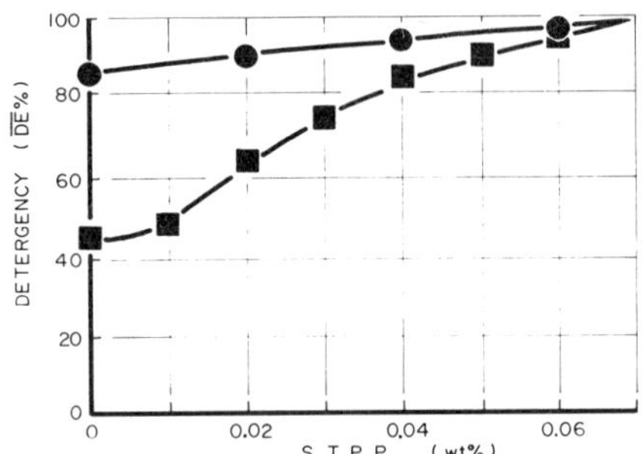

Figure 21.   Effect of STPP on detergency. Test condition: surfactant concentration—0.03 wt %. ((–●–) C_{12-14} SAE (9EO); (–■–) LAS)

Table XV   Modified "Leenerts" Test Conditions

Water Temperature	25°C
Agitation Speed	600 rpm
Wash Time	3 min.
Rinse Time	1 min.
Water Hardness	40 ppm
Detergent Solution/Beaker	700 ml
Detergent Concentration	0.04 %
Weighing	After 1.5 hr. Exposure to Air
Soil Volume/Deck Glass	23 – 27 mg
Composition of Soil Solution;	
Soybean Oil	15 parts
Tallow	15
Chloroform	100
Oil Red	0.1
Soiled Deck Glass Preparation	Deck Glasses were Immersed in Soil Solution Kept at 10±0.5°C and Dried for 12 hr.

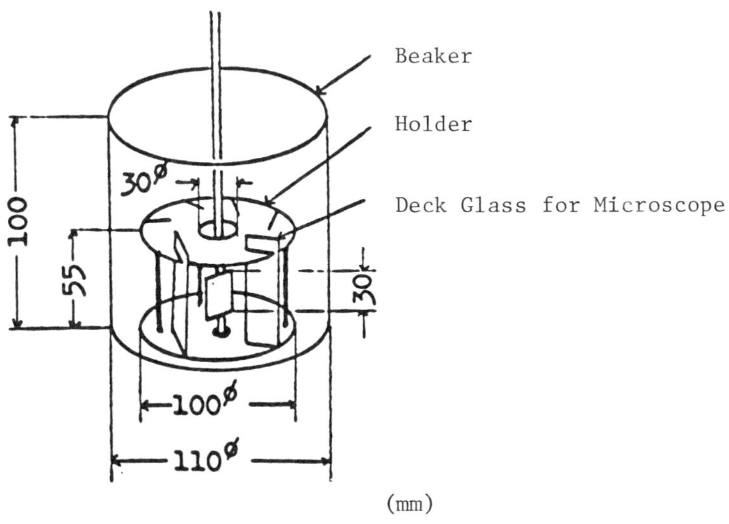

(mm)

Contrary to the case of heavy-duty detergency, the results show comparable detergencies of C_{10-12} and C_{12-14} SAE at an optimum EO mole number of about 7(Figure 22). Figure 23 shows the detergency of SAE and LAS at relatively high concentration levels. SAE(7EO), in contrast to LAS, shows detergency increasing in proportion to increased concentration. This would be of interest if applied to cleaners for heavily soiled surfaces such as kitchen ventilators. Reverse effects observed in the case of SAE(3EO) should be due to adsorption of surfactant on the soil. This means that SAE(3EO) can affect the swelling of hard oil.

Some synergistic effects were observed in binary systems consisting of LAS and SAE or SAE sulfate. Figure 24 gives the results of this study. Optimums lie between 50 and 75% of LAS in all three cases.

<u>Viscosity and Clear Point Behavior in Liquid Formulation</u>. SAE, as mentioned earlier, generally have the following advantageous characteristics, i.e. low viscosity, low pour point and high solubility. These properties lead to easy liquid formulation even at very high concentration. Flammer(<u>16</u>) pointed out this advantage in connection with the required amount of hydrotropes and concluded that SAE is far more economical a raw material for HDL than PAE.

Here, as an example, viscosity and clear point behavior of a ternary system containing LAS, C_{12-14} SAE(12EO) and water are shown in Figure 25 and Figure 26. It can easily be seen how a liquid detergent of very high concentration can be obtained.

<u>Calcium Ion Sequestering Ability</u>. To prevent eutrophication of surface water, it might be better to lower the phosphate content in detergents. On the other hand, due to the structure of washing machines and washing habits in Japan, high foam detergents are preferred from the commercial view point. In order to keep detergents high foaming, anionics are necessary. Another important point to be tested, beside foaming, is calcium ion sequestering ability. This ability was measured by the modified Murata and Arai's procedure(<u>17</u>) for C_{12-14} SAE in combination with various anionics.

The results are shown in Figure 27. It can be seen the all the anionics tested show significant effects of the combination with SAE on their sequestering ability. For this effect, the optimum EO mole number for each anionic is different. Since the combination of LAS with STPP gives a calcium ion sequestering ability of only 165 mg/g-detergents by this procedure, the combination of SAE with LAS(Linear Alkyl-benzene Sulfonate), AOS (α-Olefin Sulfonate), SAS(Secondary Alkane Sulfonate) or SDS (Sodium Dodecyl Sulfate)would be recommended for use in heavy-duty detergents.

The relationship between calcium sequestering ability and detergency or redeposition of soil is now being investigated in the authors' laboratory.

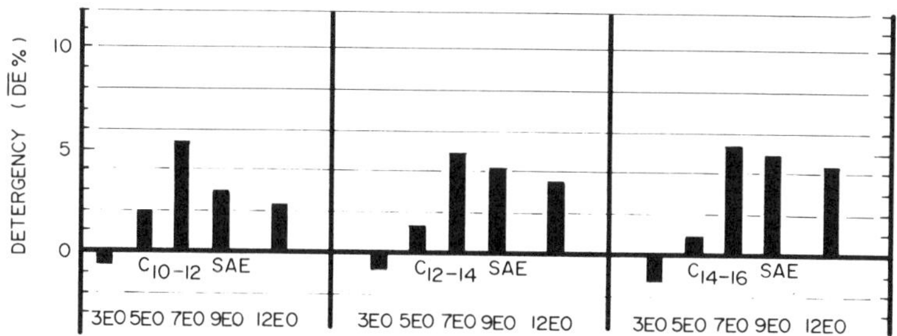

Figure 22.   Dish washing detergency of SAE in modified "Leenerts" test

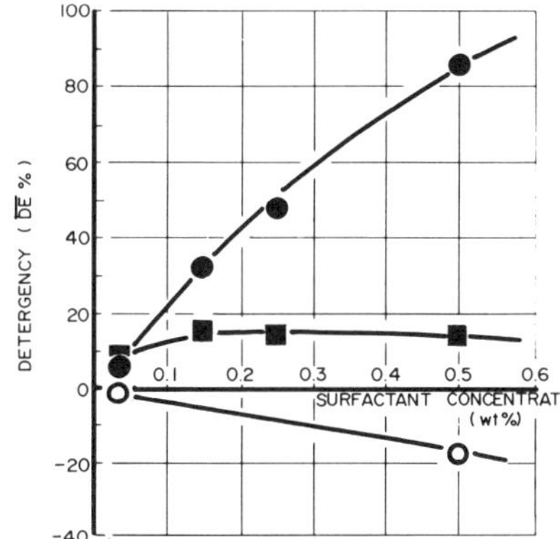

*Figure 23.   Concentration effect of surfactant in modified "Leenerts" test ((–●–)
C_{12-14} SAE (7EO); (–○–) C_{12-14} SAE (3EO); (–■–) LAS)*

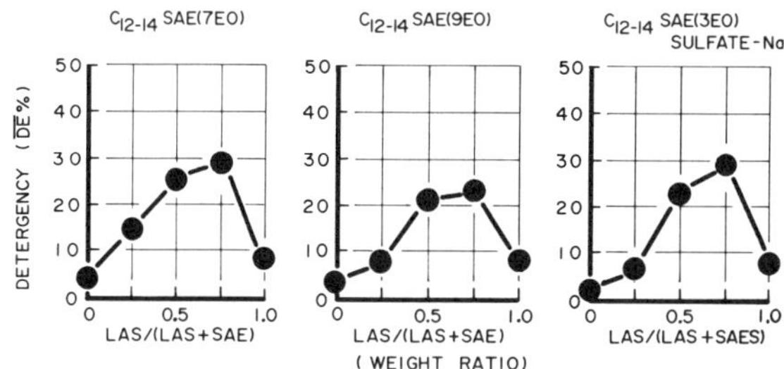

*Figure 24.   Synergistic effect on detergency in binary surfactant systems (test
method: modified "Leenerts"; condition: total surfactant concentration—0.04 wt
%)*

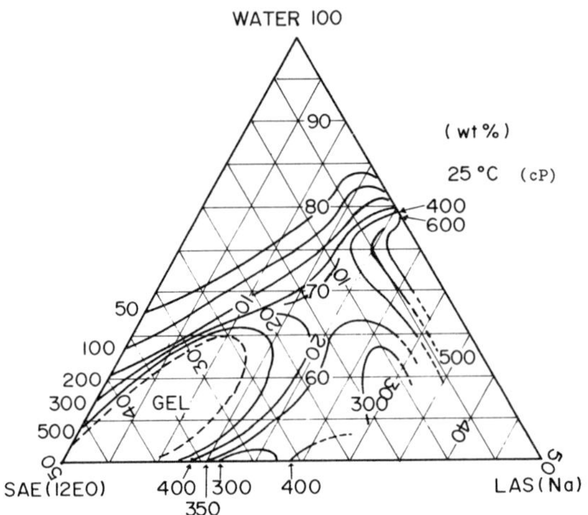

Figure 25.   Viscosity of liquid detergent based on $C_{12\text{-}14}$ *SAE(12EO), LAS(Na), and* H_2O

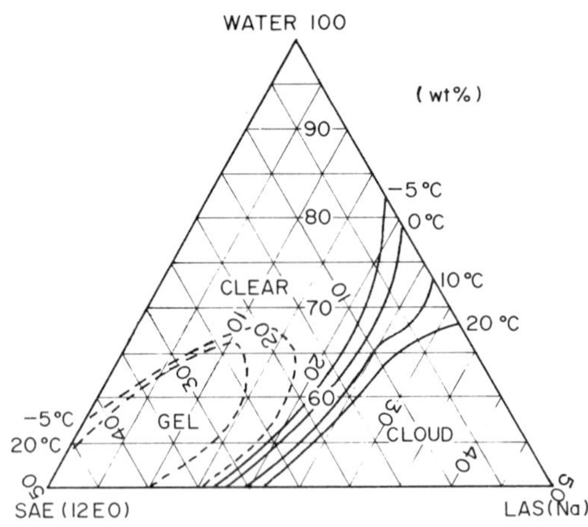

Figure 26.   Clear point of liquid detergent based on $C_{12\text{-}14}$ *SAE(12EO), LAS(Na), and* H_2O

Industrial Applications

SAE can be generally utilized as biodegradable detergents or surfactants for many industries in which the conventional alkylphenol ethoxylates have been used.

Textile Industry. Several papers dealing with applications of SAE in textile processing were already published by Zika of Union Carbide Corp.($\underline{9}$, $\underline{18}$) and Rein of Continental Oil ($\underline{19}$). Some additional performance data obtained in the authors' laboratory follow.

One example is the raw wool scouring process. The 7 mole ethoxylate of C_{12-14} secondary alcohol has excellent grease removal efficiency when used in raw wool scouring operations, as seen in Figure 28. Because of the unusually soft hand resulting, it is now used widely in Australia.

In cotton scouring, SAE also exhibits detergency comparable to that of the conventional nonylphenol ethoxylates as in Figure 29.

Not only in detergency but also in rewetting and in rinsability, SAE gives satisfactory performance. Figure 30 shows rewetting data demonstrating that SAE, over all the carbon number ranges tested, i.e. C_{10-12}, C_{12-14} and C_{14-16}, give better rewetting than nonylphenol ethoxylates.

Good rinsability of SAE has already been proven by Rein($\underline{19}$). SAE rinses from cotton, polyester/cotton and polyester/wool fabrics as easily as alkylphenol ethoxylate surfactants.

When choosing EO mole ratios of SAE for textile wet processing, one should consider the cloud point elevation at reduced surfactant concentration, which is seen in alcohol ethoxylates in general. For example, Figure 31 shows different clouding regions at different concentrations of SAE and nonylphenol ethoxylates having relatively low cloud points. The 7 mole ethoxylate of C_{12-14} secondary alcohol has its nominal cloud point at 34°C at 1% concentration. Thus, SAE(7EO) can be used at a temperature far above its nominal cloud point, giving better performance. This kind of effect is larger in SAE than in NPE.

In order to be used as an additive in various types of textile lubricants, surfactants should have high miscibility with base oils as well as the other fundamental surfactant characteristics such as lubricity improving ability and antistatic properties. Table XVI shows the miscibility of C_{12-14} SAE in various lubricant base oils in comparison with that of NPE(10EO). Good miscibility of SAE can be seen in the Table. Further information, including practical examples, is contained in Union Carbide Corp.'s booklet on Tergitol 15-S nonionics($\underline{20}$).

Some patent literature ($\underline{21}$, $\underline{22}$, $\underline{23}$) disclose that secondary alcohol esters, with or without ethylene oxide chains on the

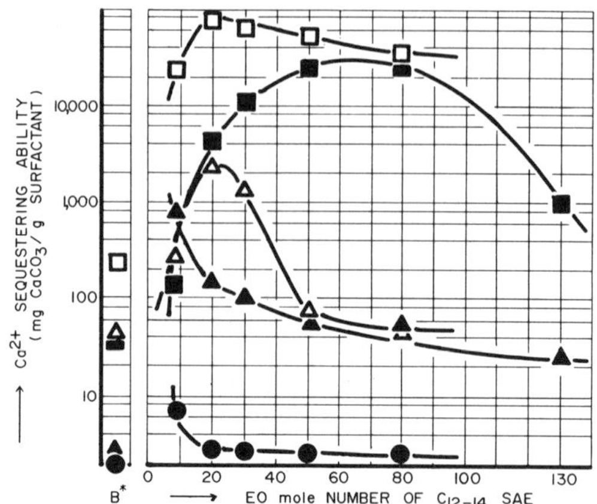

Figure 27.   Calcium ion sequestering ability of C_{12-14} secondary alcohol ethoxylates in combination with an anionic surfactant.   Test conditions: surfactant concentration: nonionics—0.1 wt %; anionics—0.1 wt %; B = without nonionics (blank).   ((-●-) with sodium laurate; (-▲-) with SDS; (-△-) with SAS; (-■-) with LAS;(-□-) with AOS)*

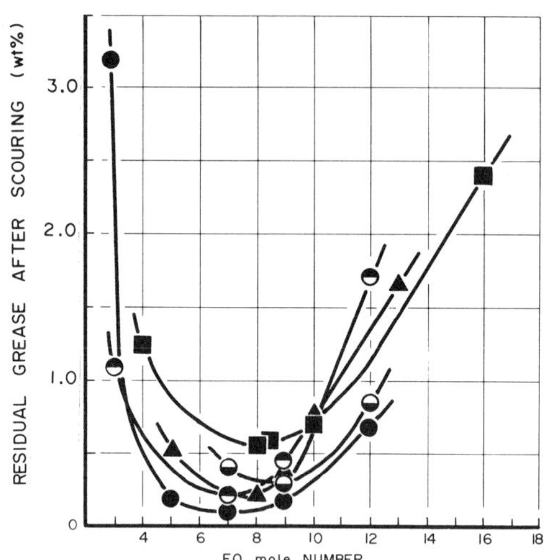

Test Conditions :

System : 4 - Bowl Scouring, with Alkaline Builder			
Stage	1	2	3	4
Liquor Volume (1)	1	1	1	1
Na_2CO_3 Conc. (wt%)	0.2	0.1	0	0
Surfactant Conc. (wt%)	0.1	0.2	0.2	0

Water Temperature   50 °C

Immersion Time   2 min. in Each Bowl

Raw Wool   AGA-79, Australia (13.9 % Grease in Unscoured)

5 g x 5 skeins

Figure 28.   Wool scouring efficiency of ethoxylates ((-◐-) C_{10-12} SAE; (-●-) C_{12-14} SAE; (-◒-) C_{14-16} SAE; (-▲-) C_{12-13} PAE (Oxo); (-■-) NPE)

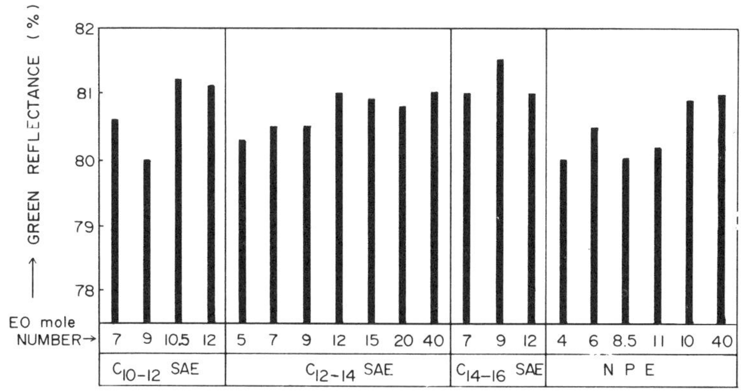

Figure 29. Cotton scouring efficiency of ethoxylates (test conditions: apparatus—Terg-O-Tometer; water temperature—80°C; agitation speed—95 rpm; scouring time—60 min; rinse time—30 min; surfactant concentration—0.2 wt %; NaOH concentration—4.0 wt %; liquor volume—1.0 L; reflectance before scouring—74.3%)

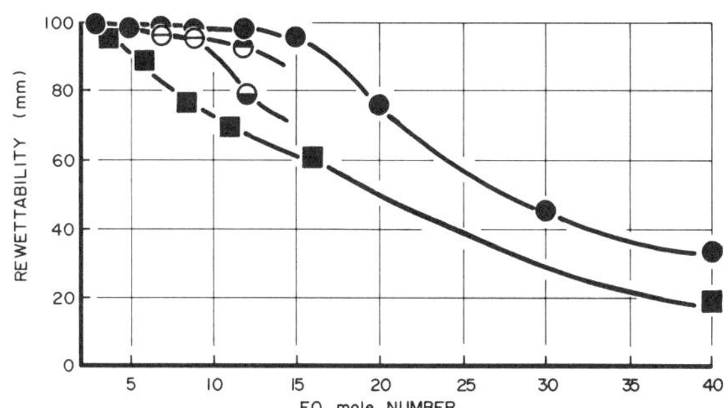

Figure 30. Rewetting data. Capillary rise method: U. Baumgarte: Melliand Textilberichte, 49, 1306 (1968). (mm): height of the elevation of test solution. ((−◓−) C_{10-12} SAE; (−●−) C_{12-14} SAE; (−◒−) C_{14-16} SAE; (−■−) NPE)

Table XVI   Miscibility of C_{12-14} Secondary Alcohol
Ethoxylates in Various Lubricant Base Oils

	$\underline{C_{12-14}}$Sec. Alcohol Ethoxylates					Nonylphenol
	3EO	5EO	7EO	9EO	12EO	10EO
(At 30°C)						
Water	C	B	A	A	A	A
Spindle Oil	A	A	A	A	C	C
Liquid Paraffin	A	B	C	C	C	C
Butyl Stearate	A	A	A	A	A	A
Methyl Oleate	A	A	A	A	A	A
Dioctyl Sebacate	A	A	A	A	A	A
Rapeseed Oil	A	A	A	A	B	C
(At 60°C)						
Coconut Oil	A	A	A	A	A	B
Tallow Oil	A	A	A	A	B	C
Triglyceryl Stearate	A	A	A	A	A	A

A = Miscible   B = Dispersed or Emulsified   C = Immiscible

Base Oil/Surfactant = 80/20

alcohol, are useful in fiber treating processes because of their
low pour points, high smoke points, low volatilities at high temperature and low coefficients of fiber to metal friction.
Typical examples are the stearate of C_{11-15} SAE(3EO) (21) and
benzoate of C_{12-14} SAE(3EO) (22).

   <u>Pulp and Paper Industry</u>. The pulp and paper industry is
another area in which SAE can be widely utilized. In the
deresination of pulp, C_{12-14} SAE are suitable biodegradable
surfactants to replace alkylphenol ethoxylates.
   Table XVII is an example of the test result showing comparable efficiency of SAE, PAE and alkylphenol ethoxylates in resin
removal from sulfite pulp. A Japanese patent specification(24)
by Lion Fats and Oils describes 10-16 mole ethoxylates of C_{8-15}
secondary alcohols as extremely useful in deresination of pulp,
especially in the manufacture of dissolved pulp for the rayon
industry. According to the patent specification, C_{12} to C_{14} alkyls
give the best results among 12 mole ethoxylates; ethoxylates of
from 10 to 13 moles of EO can give the best results among other
ethoxylates of C_{12-14} secondary alcohols, and the optimum amount
of SAE(12EO) is in the range between 0.2 and 0.6% by weight based
on pulp.
   Another Japanese patent(25) to Lion Fats and Oils claims the
use of SAE in combination with Pluronic® type nonionics in the
deresination process.
   EO/PO adducts of secondary alcohols are also attractive
surfactants in this application field. SAE and also EO/PO adducts
of secondary alcohols are efficiently used in paper machine felt
washing. In this case, their low foaming properties give
advantageous effects. They are also being used in combination
with nonylphenol ethoxylates to reduce the load on biochemical
treatment of waste effluents. A mixture consisting of 25%
SAE(12EO) and 75% NPE(10EO) also gives synergistic effects in the
deresination of pulp.

   <u>Miscellaneous</u>. Nonionic surfactants, in general, have wide
and highly fragmented application areas. Each type finds preference depending on the individual industry. SAE also have many
miscellaneous applications in which their characteristics are
advantageously utilized.
   Several interesting recent patent literature references are
illustrated as follows:

   a. Emulsifiers for agricultural chemicals(26)
      Example : C_{12-14} SAE(8EO) for emulsifying machine oil
      Effects : 1. Good surface tension lowering
                2. High wetting power
                3. Stable emulsion

b. Emulsifiers for emulsion polymerization of vinyl monomer
(27)
    Example : C_{12-18} SAE(50EO) for the emulsion polymeriza-
    tion of ethyl acrylate
    Effects : Chemically and mechanically stable emulsified
    resin products

c. Spreaders for agricultural chemicals (28)
    Example : C_{12-14} SAE(7EO)   20 parts by weight
    Oleic acid(5EO)   10   "
    Methanol   20   "
    Water   50   "
    Effects : 1. Low surface tension
    2. High wettability
    3. Sharp contact angle

d. Water soluble quenching agents for metal hardening (29)
    Example : C_{12-14} SAE(15EO)   5 parts by weight
    Water   95   "
    Effects : 1. Efficient cooling ability
    2. Small effect of concentration on cooling
    ability

SAE could be expected to be utilized in Enhanced Oil Recovery, because their low interfacial tension and low interfacial viscosity might be predicted. Finally, emulsion breakers for crude oil may be another interesting application of SAE.

Derivatives of Secondary Alcohol Ethoxylates

Propylene Oxide Adducts. SAE, in general, have better foam breakability among the nonionics, presumably due to the secondary alkyl structure and narrow distribution of EO chain lengths. Besides, modification of nonionics by the addition of propylene oxide (PO) to the terminal hydroxyl group generally reduces the foaming capacity(30). The addition of PO to SAE may, therefore, lead to nonionics with low and fast breakable foams.

The foam behavior of EO/PO adducts of secondary alcohols, primary alcohols and nonylphenols is shown in Figure 32. It can be seen that the best foam breakability is with SAE(7EO)(4.5PO), among the three.

PO adducts of SAE having various degrees of polymerization of EO and PO have been widely used in both household and industrial application areas where low and fast breaking foam is required. Textile scouring, machine felt washing in the paper industry, industrial hard surface cleaning, and machine dish washing (both in washing and in rinsing) are the main application areas so far.

Furthermore, as the combination of SAE with PO adduct of SAE gives a very low viscosity index, it can be effectively used where

Table XVII Deresination of Pulp Using Nonionic Surfactant

Test Condition : Pulp = LDSP, After Mild Alkaline Extraction, Water=64%, Resin=0.86%*, Pulp Concentration=10%, Alkali Concentration=2%, Temp.=100°C, Time=90min.

Surfactant	Surfactant Concentration* (%)	Resin Content After Treatment* (%)	Resin Removal** (%)
–	–	0.441	48.7
C_{12-14}SAE(9EO)	0.1	0.427	50.4
	0.2	0.378	56.0
	0.3	0.316	63.3
C_{12-14}SAE(12EO)	0.1	0.413	52.0
	0.2	0.335	61.0
	0.3	0.256	70.2
C_{12-15}PAE(9EO) (Oxo Alcohol)	0.1	0.434	49.5
	0.3	0.323	62.4
Nonylphenol(10EO)	0.1	0.405	52.9
	0.2	0.326	62.1
	0.3	0.236	72.6
Octylphenol(10EO)	0.3	0.298	65.3

* Based on Dry Pulp
** 100 X (0.86 – % Resin After Treatment) / 0.86

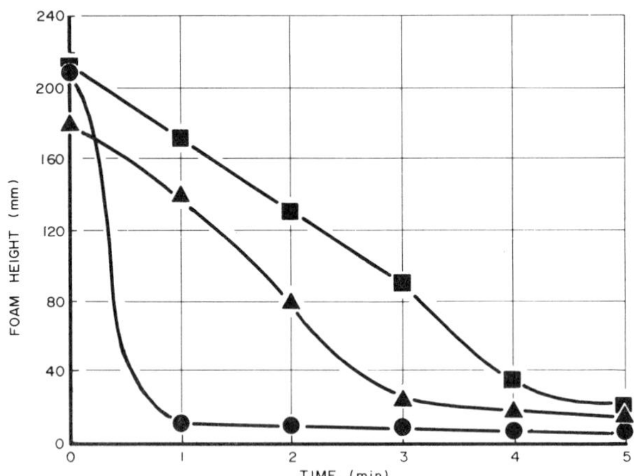

Figure 31.   Cloud point of ethoxylates at reduced concentrations

Figure 32.   Foam properties of propylene oxide adduct of secondary alcohol ethoxylate compared with those of others. Test conditions: method—Ross-Miles (JIS-K-3362); surfactant concentration—1.0 wt %; water temperature—25°C. ((−●−) C_{12-14} SAE (7EO) + (4.5PO); (−▲−) C_{12+14} PAE (7EO) + (4.5PO); (−■−) NPE (8EO) + (4.5PO))

automatic flow or volumetric control is necessary. The applicable
fields will surely be widened.

Ethoxysulfates. It is said that there are some difficulties
in producing ethoxysulfates from SAE by conventional sulfation
techniques using sulfur trioxide or chlorosulfonic acid, mainly
due to the low yield and unpleasant odor of the product (31).
Detergent formulators, therefore, have hesitated to use this
material.

Nippon Shokubai's modified ammonium sulfamate method, however,
can improve the product quality and the reaction yield. A yield
of above 99% can be obtained and the resulting product is almost
odorless with excellent stability. Skin irritation is much less
compared with the product made by chlorosulfonic acid because of
the low level of impurities found owing to the mild reactant. See
(Table XVIII).

The application possibilities have thus been widened.

Table XVIII   Skin Irritation Levels of SAE-sulfates Made by
Different Sulfonating Agents*

| Sample | Sulfonating Agent | Conc. (%) | pH | Skin Irritation | | |
| | | | | Draize's Score | | |
				Erythema & Eschar	Edema	Net Rating**
SAE sulfate(Na)	HSO_3Cl	20	7.2	2 - 8	0 - 3	C
SAE-sulfate(NH_4)	NH_2SO_3H	20	5.8	0	0 - 6	B
SAE-sulfate(Na)	"	20	6.8	0	0 - 5	B

* These data were provided by Y. Yanagimoto (The Life Science
Laboratory)
** A = None, B = Mild, C = Moderate, D = Severe. See Table XX.

Sulfosuccinate Half Esters. To meet the demand for surfact-
ants having low toxicity, especially low skin irritation, Nippon
Shokubai commercialized a series of sulfosuccinate half esters of
SAE, SOFTANOL-MES series, by it's own technology(32, 33, 34).

Figure 33 shows the synthetic route. The intermediate,
sodium maleate half ester, is relatively easily hydrolyzed in
water, while the end product, disodium sulfosuccinate half ester,
is not appreciably hydrolyzed in water in either acidic or neutral
condition (Figure 34).

The surface tension lowering effects of this type of anionic
are relatively weak, and their CMC values are a little higher than
those of ordinary anionics (Table XIX). Their foaming ability is
much higher than the original SAE (Figure 35), while their wetting
power is lower than LAS (Figure 36). Their calcium sequestering

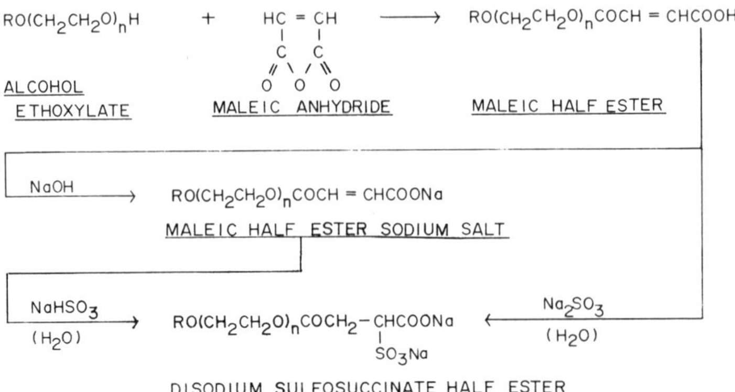

Figure 33.   Synthetic route to disodium sulfosuccinate half ester

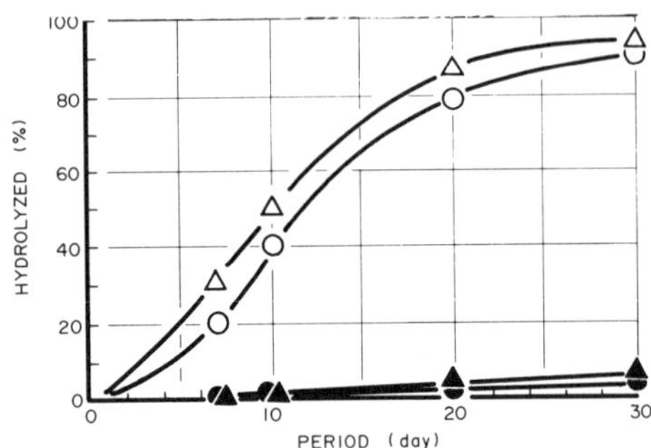

Figure 34.   Hydrolysis rate of sodium maleate half ester and disodium sulfosuccinate half ester of SAE. Test conditions: surfactant concentration—5 % aq. soln.; temperature—40°C; initial pH—7.0. ((–○–) sodium maleate half ester of C_{12-14} SAE (3EO); (–△–) sodium maleate half ester of lauryl alcohol 3EO; (–●–) disodium sulfosuccinate half ester of C_{12-14} SAE (3EO); (–▲–) disodium sulfosuccinate half ester of lauryl alcohol 3EO)

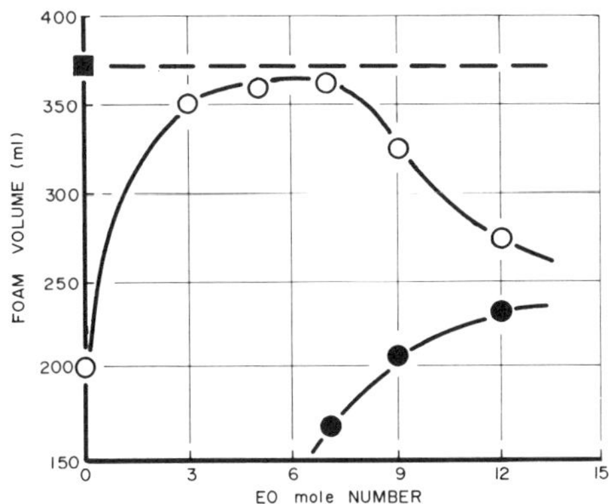

Figure 35. Initial foam volume of disodium sulfosuccinate half ester of C_{12-14} secondary alcohol ethoxylates. Test method: agitation method: surfactant concentration—0.1 %; surfactant solution—100 mL; water temperature—25°C; agitation speed—3500 rpm; agitation time—2 min. ((–○–) disodium sulfosuccinate half ester of C_{12-14} SAE series (SFT-MES-series); (–●–) C_{12-14} SAE; (–■–) LAS)

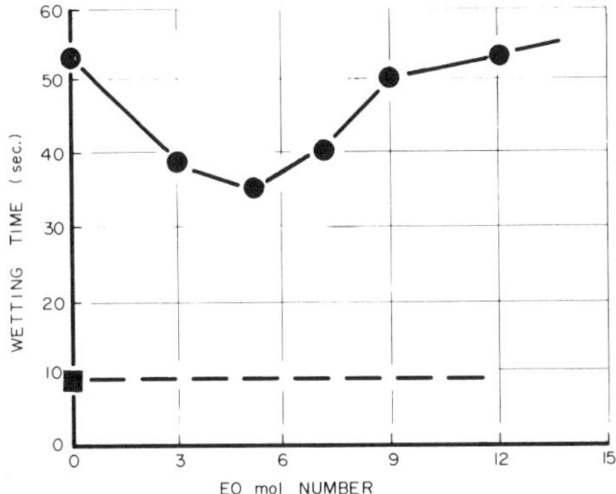

Figure 36. Wetting power of disodium sulfosuccinate half ester of C_{12-14} secondary alcohol ethoxylates (SFT-MES-series). Test conditions: surfactant concentration—0.1 wt %; cloth—wool. ((–●–) SFT-MES-series; (–■–) LAS)

ability increases with increasing EO mole number, approaching to
about the same level as the original SAE (Figure 37). This
ability of LAS, SDS, sodium laurate and other anionics is significantly increased when combined with a sulfosuccinate of SAE(Figure
38). For this effect, the optimum EO mole number for each anionic
is different.

In spite of their relatively high surface tension and poor
wetting power, detergency tests show their better performance than
LAS in non-built and comparable performance to LAS in built
systems (Figure 39).

The toxicological data for the sulfosuccinate esters are
summarized in Table XX. Products of this type in general show low
toxicity and low irritation. Eye irritation decreases with
increasing EO mole number, approaching the non-irritating level.
Further, both skin and eye irritation of AES (Sulfate of C_{12} PAE)
can be significantly weakened when mixed with the sulfosuccinate
(Table XX). This effect may suggest their applicabilities in
personal care products such as hair or body shampoo, baby soap and
medical soap.

Aquatic toxicity is also reduced when a SAE is converted to
its sulfosuccinate. Fish toxicity, TLm_{48}, of C_{12-14} SAE(7EO)
changes from 8 to 118 ppm with conversion into its sulfosuccinate.

In addition to their low toxicity and irritation, sulfo-
succinates of SAE may be recognized as anionic surfactants having
nonionic properties and performance. Their wide application,
therefore, should be considered.

Postscript

Nowadays, it is believed worldwide that n-paraffins, which
are extracted from kerosene, will be more advantageous raw material than ethylene from naphtha cracking in view of both resource
saving and process economics.

A problem to be solved for surfactant range alcohols based
on n-paraffins is how to utilize all n-paraffins in crude oil,
which have wide carbon number ranges, by finding effective
applications of the products therefrom.

As to the industrial surfactants, the shortage of sperm
alcohol, due to the restrictions on whale fishing, has increased
the demand for longer carbon chain alcohols.

In household detergents, owing to the controls and restrictions on phosphate builders, a growing demand for lime dispersing
and high detergency surfactants will necessitate increased
production of alcohol ethoxylates having relatively high carbon
number ranges.

On the other hand, as production of liquid detergents, either
heavy- or light-duty, increases, alcohols with shorter carbon
chains become important because of their good liquidities and
high solubilities. In fact, this tendency has already appeared
in some commercial products.

Table XIX Surface Tension and Critical Micelle Concentration
of Disodium Sulfosuccinate Half Ester of C_{12-14}SAE

| Surfactant | Surface Tension(dyne/cm)[*] | | | C M C | |
	0.1%	0.01%	0.001%	wt%	m.mol/kg
SFT-MES-0[**]	43.2	52.2	61.0	1.0	23
SFT-MES-3	39.0	48.4	60.0	0.20	3.6
SFT-MES-5	39.2	44.6	60.3	0.08	1.5
SFT-MES-7	39.3	45.4	61.1	0.08	1.2
SFT-MES-9	39.6	47.5	61.5	0.15	1.8
SFT-MES-12	40.9	49.0	61.9	0.20	2.1

[*] : Du Nouy Method, at 25°C
[**] : EO mole Number

Table XX Toxicological Properties of Disodium Sulfosuccinate
Half Ester of C_{12-14} SAE(SOFTANOL-MES Series)[*]

No.	Sample	Concentration (% by wt.)	pH (5%)	Irritation(Rabbit)[**] To Skin	To Eye	LD_{50}(Rat) (ml/kg.day) Male	Female
(1	SFT-MES-3[***]	20	6.5	A	C-D	17.9	18.6
(2	"	10	6.9	---	B		
(3	"	5	6.9	---	A		
(4	SFT-MES-7	20	6.8	A	B-A	40<	40<
(5	SFT-MES-12	20	6.8	A	A	40<	40<
(6	Lauryl-MES-3	20	6.8	A	C-D		
(7	Oxo-Alc.-MES-3	20	6.8	A	---		
(8	Lauryl-AES-2	20	7.3	C	C-B		
(9	C_{12-14}Sec-AES-3	20	7.2	B	B	15.2	13.7
(10	C_{12}Soap	20	9.8	C	C		
(11	L A S	20	7.7	C	---		
(12	(1 + (8 (1:1)	20	6.6	A	B-A		
(13	(4 + (8 (1:1)	20	6.7	A	A		

[*] : These data were provided by Y. Yanagimoto (The Life Science
       Laboratory)
[**] : Method, by Draize's score. (Assoc. of Food and Drug
       Officials of the U.S., Austin, TX pp. 46-59(1959)).
       Irritation Rating : A = None, B = Mild, C = Moderate,
       D = Severe.
[***] : EO mole Number.

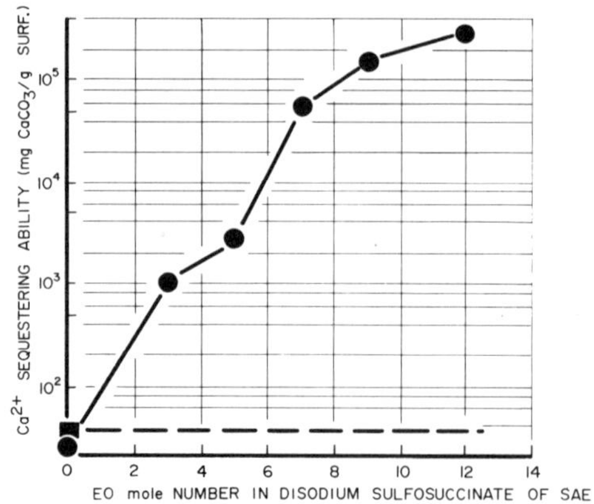

Figure 37.   Calcium ion sequestering ability of disodium sulfosuccinate half ester of C_{12-14} secondary alcohol ethoxylates (SFT-MES-series) ((–●–) SFT-MES-series; (–■–) LAS)

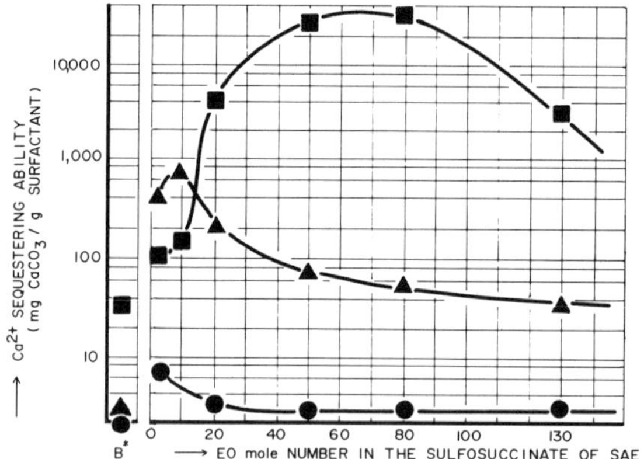

Figure 38.   Calcium ion sequestering ability of disodium sulfosuccinate half ester of C_{12-14} SAE (SFT-MES-series) in combination with another anionic. Test conditions: surfactant concentration: SFT-MES-series—0.1 wt %; others—0.1 wt %; B = without SFT-MES-series (blank). ((–●–) with sodium laurate; (–▲–) with SDS; (–■–) with LAS)*

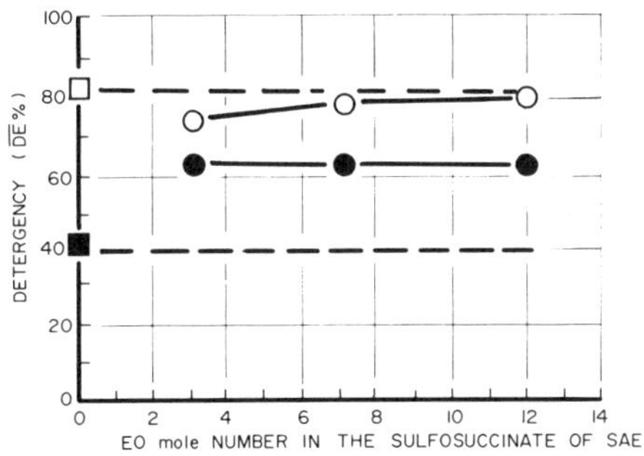

Figure 39.   Detergency of disodium sulfosuccinate half ester of C_{12-14} SAE (SFT-MES-series) with or without builder.   Test conditions: surfactant concentration—0.03 wt %; builder concentration—0.12 wt %; cloth—cotton broadcloth; builder composition—shown in Figure 20; other conditions—shown in Table XIV. ((–○–) SFT-MES-series and builder; (–□–) LAS and builder; (–●–) SFT-MES-series alone; (–■–) LAS alone)

Colin Houston & Associates has recently reported(35) that
"Such legislation as the Toxic Substances Control Act, the 1977
amendments to the Clean Water Act, and the Safe Drinking Water
Act cloud the future of the more environmentally persistent
surfactants containing cyclic rings in their structure, - - - - -
Depending on the EPA's implementation of its mandates, the
industrial use of surfactants could be altered toward the linear
surfactants synthesized from paraffins and alcohols".

Secondary alcohol ethoxylates derived from n-paraffins
should, therefore, find wide and long term demand because of
their safety and low skin irritation, and the increasing demands
for liquid and low foam detergents.

Acknowledgment

The authors thank many of their colleagues and especially
A. Nakaishi, K. Rakutani, and Y. Hayashi for many experiments
during this work.

Literature Cited

1. Kurata, N.; Koshida, K. Hydrocarbon Processing, 1978, 57, (1),
   145.
2. Kodo, N.; Kaneko, S.; Kurata, N. Yukagaku, 1975, 24, 427.
3. Helthaler, T.; Peter, E.(Riebecksche Montanwerke A.G.), Ger.
   552,886, 1928.
4. Maeda, I.; Yokoyama, H.; Kurata, N.; Okuda, Y.(Nippon Shokubai),
   Japanese, S48-37242, 1973.
5. MacFarland, J. H.; Kinkel, P. R. J. Am. Oil Chem. Soc., 1964,
   41, 742.
6. Matson, T. P. "Detergents in the Changing Scene", A Short
   Course Sponsored by The Am. Oil Chem. Soc.'s Education
   Committee, Pensylvania, 1975; p. 28-32.
7. Smith, G. D. J. Am. Oil Chem. Soc., 1979, 56, 88.
8. Fisher, C. H. ibid, 1979, 56, 918.
9. Zika, H. T. Textile Chemist & Colorist, 1969, 1, (15), 26.
10. Inoue, Z.; Fukuyama, J.; Honda, J. Mizushori Gijyutsu, 1977,
    18, 119.
11. Kurata, N.; Koshida, K. Yukagaku, 1975, 24, 879.
11a.Lashen, E. S.; and Booman, K. A. Water and Sewage Works,]967.
12. Kurata, N.; Koshida, K. Yukagaku, 1976, 25, 499.
13. Conway, R. A.; Waggy, G. T. American Dyestuff Reporter, 1966,
    Aug. 1, 33.
14. Kurata, N.; Koshida, K.; Fujii, T. Yukagaku, 1977, 26, 115.
14a.Macek, K. J.; and Krzeminski, S. F. Bulletin of Environmental
    Contamination and Toxicology, 1975, 13(3), 377-384.
15. Dillan, K. W.; Goodard, E. D.; MacKenzie, D. A. J. Am. Oil
    Chem. Soc., 1979, 56, 59.
16. Flammer, H. R. Soap/Cosmetics/Chemical Specialties, 1976, 52,
    38.

17. Murata, M.; Arai, A. J. Chem. Soc. Japan, 1974, (9), 1724.
18. Zika, H. T. J. Am. Oil Chem. Soc., 1971, 48, 273.
19. Rein, H. T. American Dyestuff Reporter, 1967, Nov. 20, 55.
20. Union Carbide Corp., Textile and Paper Chemicals, "Tergitol®
    15-S Nonionic Surfactants for Use in Textile Lubricants",
    1973, F-44456.
21. Dombrow, B. A.; Teaneck, N. J. (Diamond Shamrock),
    U.S. 3,758,594, 1971.
22. Fujimoto, T.; Uemura, T.; Ii, M. (Sanyo Chemical),
    Japanese. S53-32438, 1978.
23. Fujimoto, T.; Uemura, T.; Ii, M. (Sanyo Chemical),
    Japanese Open., S49-31995, 1974.
24. Sakuma, K.; Miyanaga, S.; Tatehara. S. (Lion Fats and Oils),
    Japanese, S50-22606, 1975.
25. Sakuma, K.; Miyanaga, S.; Koyasu, Y. (Lion Fats and Oils),
    Japanese. S53-7522, 1978.
26. Kishi, K.; Yamada, M.; Ohshima, R. (Sanyo Chemical),
    Japanese, S54-2999, 1979.
27. Yamazaki, K.; Ogata, Y.; Ishikawa, Y.;Kawaguchi, K.;
    Takeuchi, S. (Kao Soap), Japanese. S54-36951, 1979.
28. Kishi, K.; Yamada, M.; Ohshima, R. (Sanyo Chemical),
    Japanese Open. S49-30539, 1974.
29. Handa, T.; Sugiyama, K,; Tanaka, T. (Asahi Denka),
    Japanese Open. S50-98409, 1975.
30. Schick, M. J., Ed. "Nonionic Surfactants"; Marcel Dekker:
    New York, 1967; p. 411
31. Kurata, N. Separation Process Engineering, 1978, 8, (1) 29.
32. Kurata, N.; Bansho, K.; Goto, T. Japanese Open. S54-24818,
    1979.
33. Kurata, N.; Bansho, K.; Goto, T. Japanese Open. S54-24825,
    1979.
34. Kurata, N.; Bansho, K.; Goto, T. Japanese Open. S54-25286,
    1979.
35. Chemical Marketing Reporter, 1979, Nov. 5, p. 44.

RECEIVED February 2, 1981.

Higher Linear Oxo Alcohols Manufacture

R. E. VINCENT

Shell Oil Company, P.O. Box 3105, Houston, TX 77001

Most synthetic higher alcohol production today utilizes either a variation of the original Oxo technology or the Ziegler technology. This paper presents an example of the Shell process which is a modified Oxo process. This process is very flexible and with associated processes has been developed and utilized by Shell through an integrated package that yields high quality biodegradable detergent materials based on hydrocarbon distillates. This series of processes includes the following:

- Ethylene production via cracking of hydrocarbons from crude oil fractions. This process is in general use.

- Oligomerization of ethylene to higher even carbon number alpha olefins. This is the growth part of the Shell Higher Olefin Process (SHOP) Unit.

- Production of higher olefins from lower and higher linear alpha olefins via isomerization and disproportionation (I/D). This is the detergent olefin production part of the SHOP unit.

- Hydroformylation of the internal olefins from the above processes to alcohols utilizing a proprietary catalyst system. This is the part covered in this paper. It can utilize many possible feed stocks, including those from the SHOP unit.

- Ethoxylation of the alcohols to biodegradable detergent species using ethylene oxide from direct ethylene oxidation. This process is also in wide use.

The processes are flexible and efficient and the alpha olefin products from the SHOP unit can be reacted in either the SHOP I/D unit or in the hydroformylation process depending on which is most desirable. This then avoids the problem of low value byproducts.

Some aspects of these processes are covered by patents owned by Shell Oil Company (1,2,3).

The original Oxo process (or more accurately, hydroformylation process) is based on reacting a mixture of 1:1 hydrogen and carbon monoxide (synthesis gas) with an olefin to produce an aldehyde one carbon number higher than the olefin. This aldehyde is then converted to the alcohol by hydrogenation after cobalt catalyst removal. Disadvantages of the original Oxo cobalt catalyst system included a requirement of about 200 atmospheres pressure, and that the catalyst be decomposed and redissolved for recycle. Various improvements have been made in this technology by different companies. Shell's catalyst system has five primary advantages over the original Oxo catalyst system. These are:

1. It is quite stable and after crude product separation can be recycled directly to the reactor system. This eliminates the need for the catalyst decomposition and redissolving steps, thus eliminating the associated losses, operating costs, and some capital investment.

2. It rapidly achieves equilibrium isomerization of the double bond and reacts the alpha olefin to alcohol. This allows many feed source possibilities.

3. It has a high hydrogenation activity so that the main product of the reaction system is alcohol, not aldehyde. This assures low make of heavy byproducts and minimizes additional hydrogenation requirements. It also utilizes commercially available 2:1 synthesis gas.

4. It gives a high ratio of linear to iso-alcohol in the product from linear feeds. This provides both desirable detergent properties and rapid biodegradability.

5. It allows operation at lower pressure, which saves both capital investment and operating costs.

There are some precautions necessary in order to maintain good catalyst activity. Since it is a reduced metal ligand complex it is sensitive to oxidation. It also must be kept balanced for good performance and stability. This is achieved by adjusting the makeup of the catalyst components.

Process Scheme

The reaction part of the process is shown schematically in Figure 1. It consists of feeding a mixture of olefin, sythesis gas and recycle (plus makeup) catalyst to a multistage reactor system. The reactor product is cooled, degassed and fed to vacuum evaporators for crude alcohol recovery.

The crude alcohol is distilled in falling film evaporators. These units are designed to provide very low residence time and also low absolute pressures. This assures low product and catalyst decomposition losses while minimizing alcohol recycle, thereby minimizing alcohol losses to the heavy ends bleed.

The evaporator bottoms can be recycled directly to the reaction system with only a very small bleed required for rejection of catalyst decomposition products and heavy ends. The catalyst is stable under evaporator conditions.

Purification of the crude alcohol takes place in five steps. A schematic drawing is shown in Figure 2. The small amounts of ester formed are reverted with a caustic-treating plus water wash. Distillation in a light ends column removes unreacted olefins, paraffin byproduct, plus other light components present in the crude alcohol. A second distillation is used to remove trace heavy components. Hydrogenation eliminates trace unsaturation, aldehydes, etc. Filtration assures clear, solids free product.

The process scheme is not complicated and yields on olefin feed are good. The major byproduct is paraffin from olefin hydrogenation in the reactor system. However paraffin make is low. Also some olefin remains unconverted, some dimer is formed, and traces of other byproducts are made. These include acetals, esters and diols. These other losses (excluding paraffins) total less than 5%.

The equipment is almost all carbon steel. Some alloy is used, where caustic is present, to minimize the risk of stress cracking.

The paraffins made can be separated and after suitable treatment, recycled to a dehydrogenation unit such as a chlorination - dehydrochlorination unit to reform the olefin. This improves the overall yield.

Product Quality and Uses

The alcohol quality is excellent. The reaction is clean in that minimal byproducts are formed. The chemical cleanups with caustic and later with hydrogen essentially eliminate close boiling impurities. The heart cutting removes light ends and heavy end impurities. The residual hydrocarbons are separated with the light ends, and the dimers and diols with the heavy ends. Typical properties of the alcohol products are good. The alcohol purity is typically above 99%. The products are clear and water white. Acidity, unsaturation and ester content are low. Residual hydrocarbons are typically below 0.3%w. Some branched alcohol is produced but linear olefins yield about 80% linear alcohols. The carbon number distribution of the alcohol will be displaced from the olefin feed distribution by about one carbon number. Olefin feed containing up to three adjacent carbon numbers can be used; more than three carbon numbers make

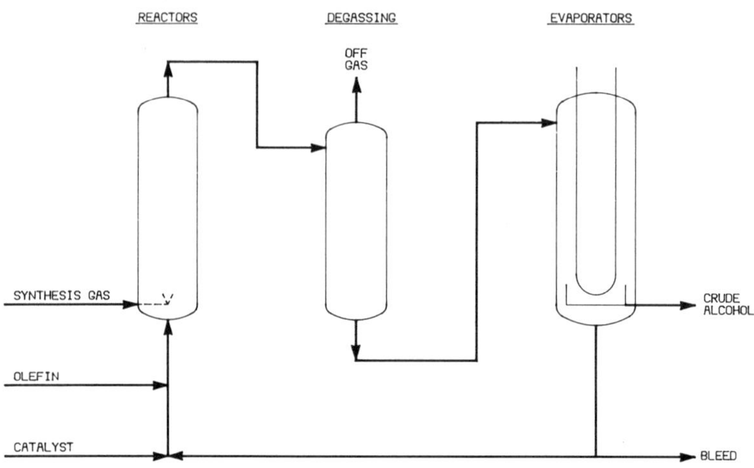

Figure 1.   Alcohol reaction and recovery system

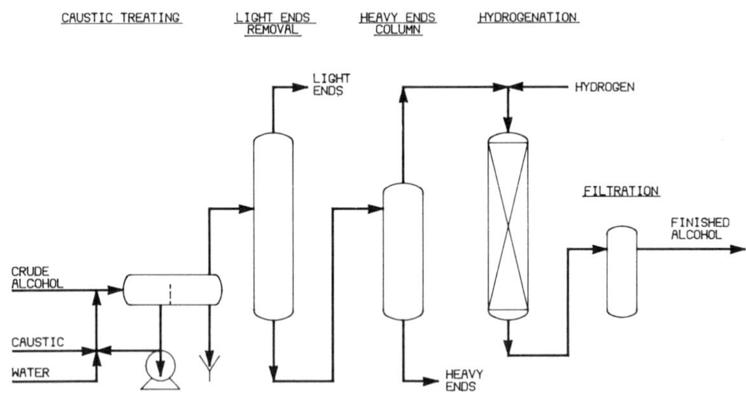

Figure 2.   Alcohol purification system

purification difficult. With narrower ranges, it is possible to switch feeds without emptying the system if the product mix can be used.

These alcohols are readily ethoxylated or sulfated to active biodegradable detergent species. The possible variations in alcohols, their mixtures and the amount of ethylene oxide or sulfation agent used provides a very wide range of possible detergent materials. Many different products are now being produced using this technology.

Feed Sources

Many types of feed have been used successfully in this process. Since the feed is narrow in boiling range and the resulting alcohol is higher boiling, the process is not sensitive to hydrocarbon feed impurities. Wax cracked olefins have been used over both the plasticizer range (C_6 to C_{10}) and the detergent range(C_{11} to C_{15}). Olefins produced by dehydrogenation of paraffins have been used. For example, both Pacol-Olex® olefins and olefins from chlorination-dehydrochlorination are acceptable feeds. The olefins from Shell's Higher Olefin Process (SHOP) are also now providing both plasticizer and detergent range alcohols. Both alpha olefins and internal olefins have been hydroformylated to alcohols. The SHOP process has been described elsewhere (4) and uses ethylene oligomerization plus isomerization and disproportionation to tailor the olefins to the desired carbon number. The ability of the hydroformylation catalyst to handle internal olefins and give essentially the same product as with alpha olefins is especially valuable with this olefin source.

Although the catalyst system produces a high ratio of linear alcohols from linear olefins, the process is not limited to linear olefin feeds. Branched olefins such as butylene dimer and propylene trimer can be used to make iso-alcohols with very little difference in yield. Also the alcohol units are designed to handle different carbon number feeds so that it is possible for a single unit to produce alcohols in the plasticizer range as well as the detergent range. This allows complete flexibility to utilize a wide range of feeds and to produce desired products.

Commercial Utilization

Shell's version of the Oxo process is in use in several foreign countries as well as in the U.S. The first commercial use of this catalyst system was for the production of normal butanol and 2-ethylhexanol in 1963. It is expected to continue to be Shell's primary process for the conversion of olefins to alcohols. This process has produced higher alcohols efficiently since its first commercial application in 1965. This is due to its simplicity, its high quality products and its flexibility to utilize many different feedstocks.

<u>Literature Cited</u>

1. Slaugh, L. H.; Mullineaux, R. D. U.S. Pat. 3,448,157 and
    3,448,158, June 3, 1969.

2. Mason, R. F. U.S. Pat. 3,737,475, June 5, 1973.

3. Verbrugge, P. A.; Heiszwolf, G. J. U.S. Pat. 3,776,975,
    Dec. 4, 1973.

4. Freitas, E. R.; Gum, C. R. <u>Chem. Eng. Prog.</u>, 1979, $\underline{75}$(1)
    73-76.

RECEIVED January 19, 1981.

Synthetic Lubricant Basestocks from Monohydric Alcohols

JACK B. BOYLAN, BRUCE J. BEIMESCH, and NICHOLAS E. SCHNUR

Emery Industries, Inc., 4900 Este Avenue, Cincinnati OH 45232

With the advent of petroleum shortages, federal Corporate Average Fuel Economy regulations and over \$1.00 per gallon gasoline, fuel economy has become the by-word of the automotive industry. As a result, lubrication along with many other automotive parameters has come under technical scrutiny as a route to improved gasoline mileage. This improved fuel economy can be achieved in crankcase oils through reducing the oil viscosity and the inclusion of friction modifiers (slippery agents) (1, 2). However, these changes must be accomplished without negatively affecting the durability or performance of the oil. The viscosity of the oil must be reduced without substantially increasing the volatility of the oil. This can be accomplished by replacing all or part of the broad distillation cut petroleum basestocks which are commonly used with narrow distillation cut synthetic basestock (3, 4). The more economic approach is to use as little synthetic as possible to accomplish your goals. Esters and polyalphaolefins appear to be excellent candidates for such an application.

It should also be mentioned that low oil viscosity at low temperatures is desirable in the new small 4 cylinder and V-6 cylinder fuel economy automobiles (5, 6). This is to permit sufficient speed of engine cranking for cold weather starting at winter temperatures. Low oil volatility must be maintained to prevent oil misting and blowby from poisoning the catalyst system and oxygen sensors in the anti-pollution devices (7).

Ester Manufacture

Most commercially manufactured esters are produced by reacting an alcohol with a carboxylic acid. This esterification reaction is reversible and is commonly referred to as a condensation reaction, since water is a by-product. The type of ester under consideration is referred to as a diester and can be produced by reacting a monohydric alcohol with a dicarboxylic acid. For example:

$$2 \text{ ROH} + \text{HO-C-(CH}_2)_x\text{-C-OH} \rightleftharpoons \text{RO-C-(CH}_2)_x\text{-C-OR} + 2\text{H}_2\text{O}$$

$$\text{Monohydric Alcohol} + \text{Dicarboxylic Acid} \rightleftharpoons \text{Diester} + \text{Water}$$

Typical raw materials used to manufacture low volatility diesters are found in Table I. An alternate route to the diesters is to start with the methyl ester of the dibasic acid and arrive at the diester via alcoholysis reaction. In this case, methanol is the co-product.

Since esterification is an equilibrium reaction, most commercial manufacturing methods generally require heating the reactants with an excess of the alcohol - the more volatile reactant. This helps drive the equilibrium in the direction of the ester product. Water of reaction is removed either by azeotroping with a special agent such as xylene or by vacuum distillation and stripping. The reaction progress is followed by measuring acidity (8). The crude ester can be refined by various methods to improve quality. For example, the ester can be treated with sodium carbonate to neutralize trace acidity (9).

Basestock Property Study

As evidenced in Table II and Figure 1, the monohydric alcohol used in making the diester has a very pronounced effect on low temperature and volatility characteristics. For comparison purposes, both petroleum and polyalphaolefin (PAO) basestocks are included in the data. The data indicate that increased branching of the alcohol has the following effects on the resulting diester:
1. Improved low temperature properties
2. Increased volatility

Increasing the molecular weight of the monomeric alcohol results in the following trends in the diester:
1. Higher viscosity
2. Lower volatility

The ideal diester for use in a partial synthetic lubricant, i.e. blend of diester, petroleum basestock and additives, should have the lowest viscosity at both high and low temperatures, and also be the least volatile. Of course, it would also need to be resistant to oxidation and corrosion and provide lubrication and wear protection when compounded into a finished lubricant. The low viscosity requirement of the synthetic portion of the partial synthetic lubricant is for economic reasons. The synthetic portion is substantially more expensive than the petroleum portion and the lower the amount required to achieve the low viscosity of the final oil, the better the final economics. As mentioned earlier, the low volatility is desirable to prevent carrying the

TABLE I

Raw Materials for Diesters

Monohydric Alcohols

No. of Carbons	Name
8	2-ethylhexyl alcohol
8	isooctyl alcohol
10	isodecyl alcohol
13	isotridecyl alcohol

Dibasic Acids

No. of Carbons	Name
6	adipic acid
9	azelaic acid
10	sebacic acid
8	phthalic anhydride

TABLE II

Basestock Properties

		Viscosity[a]		Pour Point, °F
	210°F	0°F	-65°F	ASTM D 97
di-n-octyl azelate DNOZ	3.3	Solid	Solid	+40
di-2-ethylhexyl azelate DEHZ	3.0	256	6400	-100
di-isooctyl adipate DIOA	2.4	217	5314	< -80
di-n-decyl azelate DNDZ	4.2	Solid	Solid	+75
di-isodecyl azelate DIDZ	4.3	400	24000	-95
di-isotridecyl azelate DITZ	6.6	1070	18700 (-40°F)	-70
polyalphaolefin, 6 cSt PAO-6	6.1	635	53000	-90
polyalphaolefin, 4 cSt PAO-4	4.0	347	14000	-100
100N petroleum	3.0	545	Solid	0
200N petroleum	4.4	1800	Solid	0

a   ASTM D 445 for 210°F viscosity (cSt)

   ASTM D 2602 for 0°F viscosity (cP)

   ASTM D 2532 for -65°F viscosity (cSt)

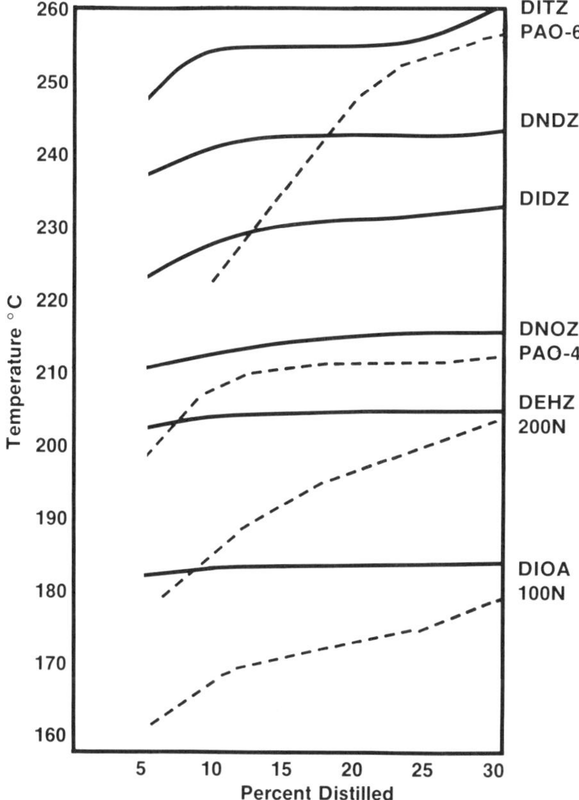

Figure 1.   ASTM D 1160 distillation (1 mm Hg)

phosphorus-containing additive to the catalyst and oxygen sensor in the anti-pollution devices and causing their premature failure through chemical poisoning (7).

Table II and Figure 1 data show di-2-ethylhexyl azelate (DEHZ) has excellent low viscosity at low temperature along with good volatility properties. Figure 1 shows the DEHZ to have substantially lower volatility than DIOA and actually lower volatility than 4 cSt polyalphaolefin (PAO-4) in the initial distilled loss area of the curves. The slopes of the PAO and petroleum curves in Figure 1 indicate they are composed of a distribution of different molecular weight species, whereas the relatively flat curves of the esters indicate they are essentially composed of a single chemical species. As a result, there are little or no low molecular weight ends to increase the volatility of these di-esters. Even though the PAO is a very narrow distillation cut, it is still composed of a blend of dimers, trimers, tetramers, and pentamers, and does exhibit volatility caused by the low ends as demonstrated in Figure 1.

Compounded Lubricant Study

It was decided to blend some of the diesters at 15% weight level into a prototype 5W-30 partial synthetic automotive crankcase formulation. The 15% level was arbitrarily assigned for economic reasons. The 0°F and 210°F viscosities were run on each formulation to define if it met 5W-30 viscosity requirements. Current SAE viscosity classifications require that a 5W-30 cross grade oil formulation exhibits a 9.6 to 12.9 cSt viscosity at 210°F and 1200 max. cP at 0°F. The latter test is referred to as a cold crank simulator test and it is supposed to correlate with low temperature cranking of an automobile engine. See Table IV for a description of the current SAE Viscosity Classification requirements. It is seen in Table III that only DEHZ, DIOA, and PAO-4 meet the 5W-30 viscosity requirements in this formulation, with the DEHZ and DIOA esters being more efficient (lower viscosity) than the 4 cSt PAO. This latter statement was confirmed by comparing the viscosity effect of different levels of DEHZ, PAO-4, and PAO-6 in 100N petroleum oil at 210°F and 0°F. The results are shown in Figures 2 and 3. In both cases, the DEHZ exhibits a superior degree of dilution efficiency.

A new proposed SAE Viscosity Classification not only reflects lower than 0°F temperature cranking characteristics of an oil, but also covers low temperature pumpability characteristics (6, 10). Low temperature cranking success without oil pumpability could result in oil starvation and subsequent engine damage. This new proposed classification attempts to establish a maximum cranking viscosity and then classifies oils according to what temperature they match this viscosity. It also includes a low shear test (mini-rotor viscometer) to duplicate pumpability of the oil at various low temperatures. Needless to say, minimum

TABLE III

Synthetic Basestock Study

| | Weight Percent | | | | | | SAE Viscosity Classifications | |
	A	B	C	D	E	F	5W	30
PMA-VI Improver	6.65	6.65	6.65	6.65	6.65	6.65		
Additives	9.10	9.10	9.10	9.10	9.10	9.10		
100SN	69.25	69.25	69.25	69.25	69.25	84.25		
DEHZ	15.00	–	–	–	–	–		
DIDA	–	15.00	–	–	–	–		
DIOA	–	–	15.00	–	–	–		
PAO-4	–	–	–	15.00	–	–		
PAO-6	–	–	–	–	15.00	–		
Vis. 0°F,cP	1082	1226	1054	1142	1294	1400	1200 max.	
Vis. 210°F, cSt	9.7	10.2	9.7	9.7	10.4	10.1		9.6–12.9
SAE 5W-30 Grade	Yes	No	Yes	Yes	No	No		

TABLE IV

SAE Viscosity Classifications

SAE Viscosity Number	Viscosity Units	At 0°F (−18°C) min.	max.	At 210°F (99°C) min.	max.
5W	Centipoises	–	1,200	–	–
	Centistokes	–	1,300	–	–
	SUS	–	6,000	–	–
10W	Centipoises	1,200	2,400	–	–
	Centistokes	1,300	2,600	–	–
	SUS	6,000	12,000	–	–
15W	Centipoises	2,400	4,800	–	–
	Centistokes	2,600	5,200	–	–
	SUS	12,000	24,000	–	–
20W	Centipoises	2,400	9,600	–	–
	Centistokes	2,600	10,500	–	–
	SUS	12,000	48,000	–	–
75W	Centipoises	150,000	@ −40°F (−40°C)	4.2 cSt	–
	SUS	–	–	40	–
80W	Centipoises	150,000	@ −15°F (−26°C)	7.0 cSt	–
	SUS	–	–	49	–
20	Centistokes	–	–	5.7	9.6
	SUS	–	–	45	58
30	Centistokes	–	–	9.6	12.9
	SUS	–	–	58	70

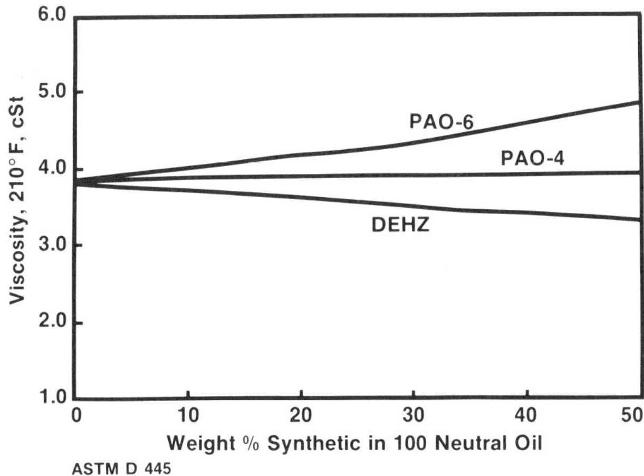

Figure 2.   Synthetic basestock blends with 100N petroleum oil

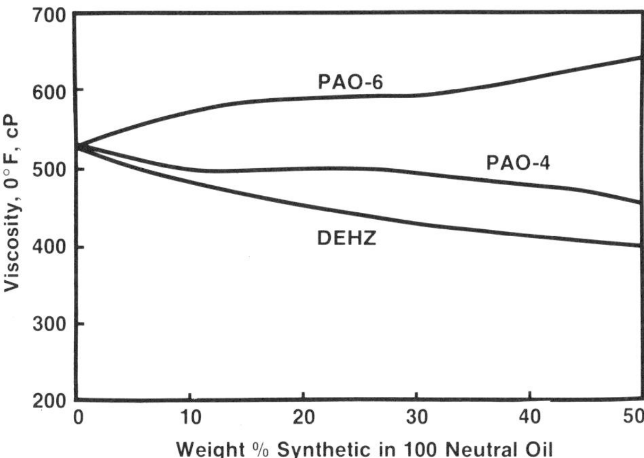

Figure 3.   Synthetic basestock blends with 100N petroleum oil

pumpability temperature should be lower than minimum crankability
temperature. These new proposed viscosity classifications are
presented in Table V. A quick review of the −65°F viscosities and
pour points in Table II will indicate that the new proposed viscosity classification chart favors the use of synthetics as viscosities below 0°F become critical.

<u>Engine Testing</u>

     Through the cooperative efforts of SAE, ASTM, and API, pas-
senger car engine oil classifications are established. These describe certain stationary engine tests which all oils must pass
before being acceptable for use in the modern day automobile engine. These requirements are continually changing as engines become more and more difficult to lubricate. Table VI describes
some of the requirements and the transition of test requirements
from classifications SE to SF. The various tests are designed to
test bearing wear and viscosity loss (L-38), rust (II-D), sludge
and varnish (III-D and V-D), and wear (III-D and VD) of the oil.
The DEHZ basestock was selected as an efficient compromise for
desired properties in a partial synthetic oil. An engine oil was
formulated using 200 neutral oil, 100 neutral oil, polymethyl-
methacrylate viscosity index improver, an additive package, and
15% DEHZ. See Table VII under Current 5W-30. This 5W-30 formulation was then subjected to the various tests described in Table VI
for qualification as an SF engine oil. The results are presented
in Table VIII.

     As can be seen, this partial synthetic passenger car crank-
case lubricant based upon the 2-ethylhexanol diester of azelaic
acid very successfully passed the engine tests required to quali-
fy as an SF grade engine oil. All of these tests were successful
on the first try. It is not unusual for multiple runs to be made
before a pass is obtained.
     On the III-D test, the 5W-30 partial synthetic showed better
resistance to viscosity increase than typical petroleum based oils.
All petroleum reference oils ran from 851% to 3077% on viscosity
increase after 64 hours. These were straight 30 weights and 10W-
30 viscosity oils as opposed to the partial synthetic being a 5W-
30 oil. Visual observation showed no scuffing of rocker arm
pivots on the first III-D test in Table VIII. Wear was observed
on seven of the rocker arm pivots. A second III-D test was run
on the slightly higher viscosity version shown under Proposed 5W-
30 in Table VII. It demonstrated excellent cam and lifter wear
with no wear or scuffing visible on any of the rocker arm pivots.
See Table VIII. The oil consumption of this formula ran 4.97
quarts versus 5.59 quarts and 5.72 quarts for the 10W-30 all
petroleum oil reference data supplied with the test.

TABLE V

Proposed SAE Low-Temperature Engine Oil Viscosity Classification

Proposed Grade	Maximum Viscosity, cP, at Temperature, °C ASTM D 2602	Maximum Viscosity (300P) and Yield Stress (105Pa) at Temperature, °C a	Minimum Viscosity cSt, at 100°C ASTM D 445
0W	3250 at −30	−35	3.8
5W	3500 at −25	−30	3.8
10W	3500 at −20	−25	4.1
15W	3500 at −15	−20	4.1
20W	4500 at −10	−15	5.6
25W	6000 at −5	−10	9.3

a   ASTM Method not Finalized

TABLE VI

<u>Current API Passenger Car Engine Oil Classification</u>

<u>Test</u>	<u>SE</u>	<u>SF</u>
II-C, rust, min. rating	8.4	–
II-D, rust, min. rating	8.5	8.5
<u>III-C</u>		
Sludge, min. rating	9.2	–
Piston Varnish, min. rating	9.1	–
Ring Land Varnish, min. rating	6.0	–
Wear-(Cam + Lifter)-avg.	0.0040"[a]	–
-max.	0.0100"[a]	–
Viscosity Increase % @ 40°C, max.	375 @ 40 hrs.	–
<u>III-D</u>		
Sludge, min. rating	9.2	9.2
Piston Varnish, min. rating	9.1	9.2
Ring Land Varnish, min. rating	4.0	4.8
Wear-(Cam + Lifter)-avg.	0.0040" max.	0.0040" max.
-max.	0.0100"	0.0080"
Viscosity Increase % @ 40°C, max.	375 @ 40 hrs.	375 @ 64 hrs.

[a]Recommended, not a required limit

<u>V-C</u>		
Sludge, min. rating	8.7	–
Piston Skirt Varnish, min. rating	7.9	–
Average Varnish, min. rating	8.0	–
Oil Ring Clogging, %, max.	5 max.	–
Oil Screen Clogging, %, max.	5 max.	–
Compression Ring Sticking	None	–
<u>V-D (9-L)</u>		
Sludge, min. rating	8.5[b]	9.4
Piston Skirt Varnish, min. rating	6.7[b]	6.6
Average Varnish, min. rating	6.3[b]	6.7
Wear-(Camshaft)-avg.	Rate & Report[b]	0.0010"
-max.	Rate & Report[b]	0.0025"

[b]Proposed as alternate to V-C limits

<u>L-38</u>		
Bearing Wt. Loss, mg., max.	40	40
Piston Varnish, min. rating	–	9.0

TABLE VII

Composition and Laboratory Test Data
On 5W-30 Partial Synthetic Motor Oils

Composition	Weight %	
Ingredients	Current 5W-30	Proposed 5W-30
100 Neutral Oil	57.25	27.25
200 Neutral Oil	12.00	42.00
PMA-VI Improver	6.65	6.65
DEHZ	15.00	15.00
Additives	9.10	9.10
Tests		
Specific Gravity	0.886	0.886
Pounds per gallon	7.38	7.38
Viscosity, cSt, 210°F	10.3	11.0
Viscosity, cSt, 100°F	57.3	62.8
Viscosity, cP, Cold Crank Simulator, 0°F	1119	1450
Viscosity Index	182	180
Flash Pt., COC, °F	415	415
Fire Pt., COC, °F	445	445
Pour Pt., °F	-30	-30
Total Base Number	1.9	1.9

TABLE VIII

Engine Test Data on 5W-30
Partial Synthetic Motor Oil

Test	Results	SE	SF
II-D, Rust	8.89	8.5 min.	8.5 min.
III-D			
(current 5W-30 visc.)			
Sludge	9.54	9.2 min.	9.2 min.
Piston Varnish	9.44	9.1 min.	9.2 min.
Ring Land Varnish	7.12	4.0 min.	4.8 min.
Wear-			
(Cam + Lifter)-avg.	0.0039"	0.0040" max.[a]	0.0040" max.
-max.	0.0046"	0.0100"[a]	0.0080"
Viscosity Increase	99 @ 40 hrs.	375 max. @	375 max. @
(% at 40°C)	136 @ 64 hrs.	40 hrs.	64 hrs.
Oil Consumption	5.63 Qts.	–	–
III-D			
(proposed 5W-30 visc.)			
Sludge	9.2	9.2 min.	9.2 min.
Piston Varnish	9.15	9.1 min.	9.2 min.
Ring Land Varnish	6.60	4.0 min.	4.8 min.
Wear			
(Cam + Lifter)-avg.	0.0026"	0.0040" max.[a]	0.0040" max.
-max.	0.0037"	0.0100"[a]	0.0080"
Viscosity Increase	59 @ 40 hrs.	375 max. @	375 max. @
(% at 40°C)	98 @ 64 hrs.	40 hrs.	64 hrs.
Oil Consumption	4.97 Qts.	–	–

[a]Recommended, not a required limit

Test	Results	SE	SF
V-D (9-L)			
Sludge	9.41	8.5 min.[b]	9.4 min.
Piston Varnish	8.10	6.7 min.[b]	6.6 min.
Average Varnish	7.92	6.3 min.[b]	6.7 min.
Wear-(Camshaft)-avg.	.0005"	Rate & Report[b]	0.0010" max.
-max.	.0008"	Rate & Report[b]	0.0025"
Oil Ring Clogging, %	0		
Oil Screen Clogging, %	0		
Compression Ring			
Sticking	0		

[b]Proposed as alternate to V-C limits

Test	Results	SE	SF
L-38			
Bearing Wt. Loss, mg.	21.9	40 max.	40 max.
Piston Varnish	9.6	–	9.0 min.

Conclusions

Diesters prepared from specific monohydric alcohols can be
formulated into low viscosity partial synthetic automotive crankcase lubricants. These diester basestocks exhibit definite advantages over petroleum basestocks of similar and sometimes
higher viscosities. These advantages are:
  1.   Much lower volatility at equal or lower viscosity at
       210°F as demonstrated by comparing DEHZ with 100N and
       200N petroleum oils.
  2.   Much improved low temperature properties at comparable
       210°F viscosity as demonstrated by comparing DIDZ with
       200N oil.
The diesters also show some advantage over PAO's in viscosity
dilution efficiency as shown in Table III where DEHZ and DIOA
gave lower 0°F viscosities than PAO-4. Figures 2 and 3 also
demonstrate this advantage when DEHZ is compared to PAO-4 and PAO-
6 in viscosity dilution effects.
The actual engine test data shown in Table VIII indicate that
diester based 5W-30 partial synthetic oil can very successfully
pass the requirements for SE/SF qualification while showing oil
consumption rates better than 100% petroleum 10W-30 oils.

Literature Cited

1. Retzloff, J. B.; Davis, B. T.; Gluckstein, M. E.; and
    Pietras, J. J. ASLE Preprint No 79-AM-2C-1, May 1979,
    "Fuel Economy Benefits from Modified Crankcase Lubricants".

2. Marshall, E. F. CLC Report No. 502, December 1978, "Survey
    of Lubricant Influence on Light-Duty Vehicle Fuel Economy".

3. Imparato, L.; Berti, F.; Mancini, G.; and Pusateri, G.
    SAE Paper 740118, February 1974, "Use of Synthetic Lubri-
    cants in Multigrade Motor Oils".

4. Campen, M.; and Myers, E. E. NPRA Meeting, November 1978,
    "Performance of Part Synthetic Engine Oils".

5. Caracciola, F.; and McMillan, M. L. SAE Paper 790728, June
    1979, "Effect of Engine Oil Viscosity on Low-Temperature
    Cranking, Starting, and Fuel Economy".

6. Stover, W. H. SAE Paper 790729, June 1979, "Cold Starting
    and Oil Pumpability - An Evaluation of New and Used Oils in
    Gasoline Engines".

7. Caracciola, F; and Spearot, J. A. SAE Paper 790941, October
    1979, "Engine Oil Additive Effects on the Deterioration of
    a Stoichiometric Emissions Control (C-4) System".

8. Gunderson, R. C.; and Hart, A. W. "Synthetic Lubricants";
    Rheinhold Publishing Corporation, 1962.

9. Chao, T. S.; Kjonaas, J.; and Dijovine, J. ACS Symposium
    on Chemistry of Synthetic Lubricants and Additives, September
    1979, "Esters From Branched-Chain Acids and Neopentylpolyols
    and Phenols as Basic Fluids for Synthetic Fluids".

10. Schaub, H.; Smith, M. F.; and Murphy, C. K. SAE Paper 790732,
    June 1979, "Predicting Low Temperature Engine Oil Pumpability
    With the Mini-Rotary Viscometer".

RECEIVED December 29, 1980.

Monohydric Alcohol Ester Plasticizers for Polyvinyl Chloride

Past, Present, and Future

E. J. WICKSON

Exxon Chemical Company, P.O. Box 241, Baton Rouge, LA 70821

As noted in the Preface, monohydric alcohols are generally categorized as lower (C_1-C_5) and higher (C_{6+}) alcohols. The largest outlet by far for higher monohydric alcohols is in ester type plasticizers for polyvinyl chloride (PVC). The PVC plasticizer range alcohols are generally considered to be C_6-C_{11}, although branched chain tridecyl alcohol is also of importance. Of the lower alcohols, only normal butanol, and to a far lesser extent isobutanol, are used in this application. Total volume of these lower alcohols used is small by comparison.

The Decade of the Thirties - The Beginning

The use of monohydric alcohol esters as PVC plasticizers had its beginning in the decade of the thirties. Plasticization of PVC was first demonstrated by W. D. Semon of B. F. Goodrich(1). Semon's patent gave examples of plasticizers such as dibutyl phthalate and other unrelated compounds. However, the claims were limited to non specific aromatic substances, liquid aromatic nitro compounds, and specific aromatic nitro compounds. The first commercially used plasticizers for PVC were those already available and used as cellulosic plasticizers, such as the monohydric alcohol ester--dibutyl phthalate.

Di-2-ethylhexyl phthalate (DOP) had already been patented as a composition of matter and examples of its use as a plasticizer for cellulosics, etc., but not PVC, were cited(2). However, it was not until later in the decade that the overall advantages of DOP were recognized. This was a key factor in the large scale production of 2-ethylhexanol (2EH) by Union Carbide Corp. starting in 1939(3), followed shortly by the commercialization of DOP -- also by Union Carbide(4). 2-Ethylhexanol was made at that time via the "double aldol" route, starting with acetaldehyde.

It is interesting to note that DOP showed no particular virtue as a plasticizer for the other resins available at the time. Were it not for flexible PVC, DOP might have passed into obscurity(3). Instead, as is well known, DOP became the industry

standard for PVC and is still the single largest volume plasticizer produced to this day, on a worldwide basis.

The Decade of the Forties - Flexible PVC Takes Off

During the early forties flexible PVC use was mainly limited
to military applications, including wire and cable, and various
film and sheeting applications, etc. With the conclusion of
World War II, new markets for PVC developed rapidly. Aside from
DOP, which was commercialized in 1940(4), dicapryl phthalate
(DCP) was introduced by Rohm and Haas(5), also in the early
forties. With the development of plastisols(6), DCP became a
plasticizer of choice because of the favorable balance of
properties it imparted -- especially low viscosity. However,
DCP's subsequent growth has been limited since capryl alcohol
(octanol-2), a monohydric secondary alcohol, is a by-product of
limited volume sebacic acid manufacture.

It is noteworthy that the only monohydric secondary alcohols
that have found any use in PVC plasticizers are capryl alcohol
(octanol-2) and cyclohexanol. No tertiary alcohols have found
use at all, probably because of the difficulty of esterification,
at least.

Other 2-ethylhexanol based plasticizers were introduced,
including some which imparted outstanding low temperature flexi-
bility--dioctyl adipate (DOA), dioctyl azelate (DOZ), dioctyl
sebacate (DOS), and trioctyl phosphate (TOF). In addition, TOF
showed high resistance to microorganisms which was important in
military applications. Furthermore, TOF improved flame resistance. However, the mixed ester--octyl diphenyl phosphate--also
introduced in the forties--was far superior and showed a better
balance of low temperature performance and flame resistance than
either TOF or the well-established plasticizer--tricresyl
phosphate (TCP). Each of these found a market as a specialty
plasticizer because of these specific performance attributes.
None, however, was a serious threat to DOP on an overall price-
performance basis. Rather, they were used to supplement the
properties of DOP where its performance was inadequate.

Meanwhile, Rohm and Haas introduced the first of the polymerics - polyesters based on the reaction of a dibasic acid with
a glycol and with a monohydric alcohol or monobasic acid as chain
terminator(5). These plasticizers were developed as highly
permanent, extraction and migration resistant types for specialty
applications. 2-Ethylhexanol and isodecyl alcohol are examples
of monohydric alcohols used as chain terminators in today's
polymerics.

The oxo process, discovered in Germany by Otto Roelen, was a
breakthrough method of producing primary monohydric alcohols.
The oxo process was commercialized by Exxon Chemical Co. in the
U.S. in 1948 with the manufacture of isooctyl alcohol(7). This

quickly led to the introduction of diisooctyl phthalate and other
ester-type plasticizers similar to those based on 2-ethylhexanol.

Still in the forties, butyl benzyl phthalate (BBP) was
introduced by Monsanto Co. This ester is based on the monohydric
alcohol n-butanol, but the benzyl group is derived from benzyl
chloride rather then benzyl alcohol. This route was selected to
make possible manufacture of pure coester which showed a better
balance of properties than if made from the two alcohols. BBP
was destined to make its mark in PVC, particularly in the
flooring industry.

Aside from the plasticizers mentioned, there was feverish
activity to develop other basically new types by a host of companies. About twenty thousand plasticizers were mentioned in
the literature by 1943(8). Certainly hundreds of candidate
plasticizers were actually sampled by the industry during the
period 1945-1955. Most, however, failed to show an advantageous
balance of performance and price when compared to already existing products.

The Decade of the Fifties - The Oxo Process Exploited

During the fifties there was widespread activity in the oxo
process area. Additional branched chain primary monohydric alcohols - isodecyl and tridecyl - were introduced by Exxon Chemical
Company. The phthalates of these alcohols were found to be particularly useful in high temperature PVC insulated wire and cable
where DTDP, for example, partially replaced the more costly
polymerics(9).

Meanwhile, lower monohydric alcohols were first made by the
oxo process. These included n-propanol, n- and isobutanol and
amyl alcohol. In PVC, only one of these assumed any importance -
n-butanol - mainly in butyl benzyl phthalate. Dibutyl phthalate
(DBP) was by this time considered too volatile for use in PVC in
the U.S.(10). The so-called coesters - butyl octyl phthalate
(BOP), butyl isodecyl phthalate and butyl cyclohexyl phthalate -
have been on and off the market for many years but never achieved
major success in the U.S. Dibutyl phthalate/DOP blends and BOP
have probably been most successful in Japan. Diisobutyl phthalate is used in Brazil and other South American countries even
today, because of its low cost. BOP is also offered in Europe,
but as in the U.S., it has not been a major success. One reason
could be the wide variability of composition from different
suppliers. In fact, some BOP's were merely physical blends of
DBP and DOP.

While n-butanol from the oxo process did not become a major
intermediate for PVC plasticizers, the development was nonetheless
a breakthrough of far-reaching significance. The precursor of
n-butanol in the oxo process is n-butyraldehyde. This made
possible the manufacture of 2-ethylhexanol by "single aldol" at
lower cost than the "double aldol" route via acetaldehyde.

The commercialization of isophthalic acid by Chevron in 1956(11) led to the introduction of di-2-ethylhexyl isophthalate (DOIP). This plasticizer showed improved lacquer mar resistance in flexible PVC(12). This need was limited, however. Because of its premium price, DOIP has not gained acceptance as a general purpose plasticizer.

Following the successful development of epoxidized soybean oil(5), mainly as a stabilizer adjuvant with high permanence but poor low temperature plasticizing properties, the epoxidized fatty acid esters were introduced. These included the C_8 monohydric alcohol esters -- octyl epoxystearate and octyl epoxytallate. They also acted as stabilizer adjuvants but with outstanding low temperature plasticizing properties. The epoxy stabilizer/plasticizers have grown to over 50,000 tons/year in the U.S. in 1978, with the epoxidized soybean oil type predominating.

The Decade of the Sixties - The Straight Chain/Linears Emerge

Hexyl alcohol, a mixture of n- and iso C_6 alcohols made via the oxo process by Exxon Chemical Co., started the flow of new monohydric alcohols as intermediates for PVC plasticizers in the sixties. Dihexyl phthalate - the phthalate of this alcohol - is a fast solvator and competes with other fast solvating types in markets such as vinyl flooring and carpet backing.

The commercialization of the Ziegler process by Conoco and Ethyl (See Chapter 7) was the next breakthrough in monohydric alcohols production. These straight-chain alcohols are derived from ethylene growth on triethyl aluminum, producing a range of even number primary monohydric alcohols. Phthalates of blends of $n-C_6-C_8-C_{10}$ and $n-C_8-C_{10}$ are known as 610P and 810P, respectively. Their lower volatility and improved economics resulted in a takeover from DOP/dioctyl adipate and similar blends. This was especially important in the automotive upholstery market where important requirements are low temperature flexibility and low fogging - i.e., plasticizers that resist vaporization from vinyl with subsequent condensation on the windshield. Actually, the straight chain ester - di-n-octyl phthalate - was available in the forties. However, n-octanol was produced from coconut oil and was not economically attractive at that time.

Another important development in the early sixties was the commercialization of trimellitic anhydride (TMA). Monohydric alcohol triesters of TMA were shown to have outstandingly low volatility while overcoming the poor low temperature properties of polymeric-type plasticizers. In addition, the TMA esters were lower in cost than the polymerics - particularly when volume cost was considered.

By the middle of the decade plasticizer production had passed the billion pound level(13), and reached 1.2 billion pounds in 1968, more than fulfilling a prediction made in 1955 by Skeist(14). More than 80% of this production was for PVC.

The next development in plasticizer range monohydric alcohols in the sixties was isononyl alcohol. This branched chain oxo alcohol and its ester (DINP) were commercialized in 1968 by Exxon Chemical Co. DINP offered better overall performance than DOP for most uses, as well as significant compound volume cost advantages. As a result, it has grown rapidly in the marketplace as a general purpose (GP) plasticizer.

As the decade ended, a new monohydric alcohol phthalate based on mixed C_7-C_9-C_{11} straight chain and single methyl branched alcohols was introduced by Monsanto Co. The alcohols are produced by the oxo process from straight chain alpha olefins. The phthalate competed directly with 610P, especially in the automotive upholstery market where it gained rapid acceptance. The term "linear" has been adopted to describe largely straight-chain alcohols(15). In fact, "linear" has come to mean straight chain or largely straight chain, particularly when describing phthalates.

The Decade of the Seventies – From Excess Capacity to Shortages

As the seventies got underway, the premium price on the GP linear phthalates decreased to where these plasticizers also began to capture a share of the general purpose plasticizer market, particularly if improved low temperature flexibility was important. The phthalate of the C_{11} fraction of the linear C_7-C_9-C_{11} alcohols was introduced by Monsanto and found acceptance in high temperature wire and cable where it competes with branched chain DTDP. In Europe, linear alcohol phthalates based on alcohol blends made by oxonation of cracked wax olefins gained acceptance. These differed from the U.S. linear alcohol blends in having both odd and even carbon chain lengths.

The next breakthrough of importance for future 2-ethylhexanol plants occurred in the mid seventies. This was the development of the rhodium-catalyzed oxo process by Union Carbide, Davy Powergas and Johnson-Matthey (See Chapter 6). This process not only operates at lower temperatures and pressures than the conventional cobalt-catalyzed process but also gives a far lower yield of the less valuable isobutyraldehyde by-product. The net result is improved economics vs. the cobalt process for n-butyraldehyde – the intermediate for 2-ethylhexanol. Although outside the U.S. this new technology has already been licensed and plants are now operating(16), no new plants were constructed in the U.S. specifically for 2EH manufacture in the seventies. However, Union Carbide plants in Texas and Puerto Rico have been modified, utilizing this new technology for propionaldehyde and butyralde-

hyde manufacture, respectively. Some of the latter is believed
to be used to make 2EH.

     Also in the mid seventies, a new type plasticizer - dioctyl
terephthalate (DOTP) - was introduced by Eastman. DOTP is less
volatile and has better low temperature flexibility than DOP,
although it is less solvating than DOP or even DIDP. A branched
chain C_9 terephthalate is reported to be compatible with PVC but
not a 70% linear C9 average phthalate. Most C_6 and lower tere-
phthalates are reported to be incompatible solids(17).

     Throughout the history of flexible PVC there have been
periods where plasticizers were alternately in a long or short
position. Looking back, the seventies was a particularly rough
period in this respect for the industry. At the beginning of the
decade the market was grossly oversupplied in terms of both
alcohol and ester capacity. This condition led to poor profita-
bility and ultimately a restructuring of the marketplace. Many
small producers not basic in alcohols and phthalic anhydride
dropped out. One major plasticizer supplier left the GP plasticizer market; another left the plasticizer market entirely, and
late in the decade a third supplier shut down its 2-ethylhexanol
plant due to lack of propylene feedstock. The decade ended with
a tight supply situation on plasticizer range monohydric alcohols
due largely to the shutdown of the major 2EH plant. Startup of
Exxon Chemical's 50% expansion in oxo alcohol capacity helped
ease what could have been a major shortage.

The Decade of the Eighties - Slower Growth Seems Probable

     At the start of the eighties the plasticizer alcohol supply
situation continued tight. However, by the end of the first
quarter of 1980 there was a sharp drop in demand which is
expected to slowly rise to 1979 levels by about 1983. Projections are for an average growth rate of about three percent for
the rest of the decade. This is largely the result of projected
reduced demand in both automotive and housing - key indicators
for the flexible vinyl industry. There are some new uses
emerging, such as vinyl roofing for industrial buildings and
automotive undercoatings - both long established in Europe - but
there are also inroads on vinyl, for example, in upholstery, wire
and cable, and shoes.

     New alcohol capacity has been announced by Monsanto Co.(18),
and a new joint venture on 2EH by Tenneco Chemicals and USS
Chemicals is reported to be approved(19). This new capacity is
not due on stream before 1982-1983, however. Also, restart of
the former Oxochem 2EH plant in Puerto Rico has been rumored, but
no definitive announcement has been made.

     Phthalates of C_8-C_{10} average carbon number monohydric
alcohols continue to dominate the PVC plasticizer market, and the
likely scenario is for this to continue through the eighties.
There have been, however, numerous developments which will

influence the relative importance of the specific monohydric
alcohols which will be used. Most of these involve economics -
of feedstocks, alcohol processes, and vinyl compounding and
fabrication. The remainder involve new or more restrictive end
use requirements. Some of these key developments are as follows:
     1. The Changing Value of Olefin Feedstocks for Alcohols.
Lower olefins - ethylene and propylene - are the basic feedstocks
for all important monohydric alcohols used in PVC plasticizer
manufacture in the U.S. Thus, the relative values of these olefins are of key importance to manufacturing cost.

     While the eighties started with a very tight supply situa-
tion on propylene, availability is not now a problem. Longer
range, the continuing trend to naphtha and gas oil as cracker
feeds in a market that is basically ethylene driven will mean
increased amounts of propylene relative to ethylene will be
produced. This is expected directionally to increase the
historical price differential between ethylene and propylene. In
addition, butylenes, which were of relatively minor importance to
higher alcohols production in the seventies, will assume increasing importance in the eighties based on announced plans of
Monsanto(18). Also, in Japan a new DINP based on Dimersol®
octenes from n-butenes is slated for commercialization in 1980-
1981(20).

     Another factor which will bear on the relative values of
these olefins is the burgeoning demand for isobutylene for methyl
t-butyl ether as a gasoline octane improver(21). Large amounts
of n-butenes will become available which in the past could be
utilized as dehydro feed for butadiene. This outlet for n-
butenes, however, is declining due to increasing amounts of
butadiene from steam crackers. Although new outlets for n-
butenes such as polyethylene copolymers are expected to grow,
this will not balance the loss of the dehydro feed outlet. The
value of n-butenes will continue to be governed by fuel products
outlets and could be low compared to ethylene and propylene.

     In view of the foregoing, it suffices to say that the rela-
tive economics of producing the various plasticizer range monohydric alcohols will be different in the eighties than in the
seventies.

     2. The Declining Market for Linear Phthalates in Automotive
Upholstery. The major market for linear phthalates has historically been automotive upholstery. However, the U.S. built share
of the automotive market has continued to decline, and the trend
(now a landslide) to smaller cars has resulted in a sharp
decrease in demand for domestic vinyl upholstery. Besides this,
fashion trends have turned to increased use of non-vinyl upholstery. Finally, the use of softer vinyls with higher plasticizer
levels means increased opportunities in the automotive upholstery
market for diisodecyl phthalate (DIDP). DIDP meets the low temperature requirements of these new softer vinyls, and at the same
time offers reduced windshield fogging and lower cost on a

compound volume basis than the GP linear phthalates. The lower
cost results from DIDP's lower efficiency and specific gravity vs.
these standard linear phthalates. The net result is that the
linear phthalates are likely to decline in relative importance
in the eighties.

    3. <u>The Increasing Acceptance of Plasticizers with Better
Cost-Performance than DOP</u>. While the breakthrough n-butyraldehyde technology of the mid seventies will tend to make new 2EH
capacity competitive with existing plasticizer range alcohol
plants, there has been a trend away from DOP in the U.S.
Reasons for this include -

    o The fogging problem in automotive interior applications.

    o New or more demanding applications beyond the capabilities
      of DOP, such as industrial vinyl roofing, and the trend
      to thinner wall wire insulation and jacketing.

    o The decline in 2EH capacity has, of course, not helped
      DOP usage. The 2EH supply situation is expected to be
      corrected in the next few years, as noted earlier.
      However, during the period when DOP was in short supply
      many users converted to DINP, DIDP, and linear phthalates.
      They have found the better overall performance, and in
      many cases better compound economics of these plasti-
      cizers justifies continued use. Although DOP will regain
      some of its lost ground, it is not expected to dominate
      the U.S. market as in the past.

    4. <u>The Increasing Cost of Energy</u>. Much attention has been
given in the literature to the use of various highly aromatic
ester-type fast solvating plasticizers as a means of reducing
energy requirements and/or increasing plant output. There are
certainly some vinyl fabrication operations where this applies.
However, increased raw material costs due to the higher cost per
pound of most of these fast solvating plasticizers and their high
specific gravities (typically 1.12-1.13) must be weighed against
possible energy savings. A new non-premium, fast-solvating
plasticizer - diisoheptyl phthalate (Jayflex® 77) - was recently
introduced by Exxon Chemical Co. This plasticizer overcomes the
high specific gravity and poor low temperature performance of the
highly aromatic fast solvating type plasticizers. Similar esters
have been available in Europe and Japan for many years and have
achieved some importance in the latter country.

    The specific gravity difference between DOP and a typical
highly aromatic fast solvator represents a volume difference of
almost 14 percent for the same weight. An example of how this
high specific gravity increases raw material cost is shown in
Table I. It is apparent that replacing 20% of the DOP in this
formulation with a fast-solvating plasticizer has increased the
raw material cost of a 100 yard roll of 54 inch width 12 mil
sheeting by $0.64. Typical energy costs for a one and one half
meter width calender producing 12 mil average thickness sheeting
are estimated at 2¢ per pound of compound(<u>22</u>). This includes

TABLE I

AFFECT OF HIGHLY AROMATIC FAST SOLVATOR
ON PVC COMPOUND RAW MATERIAL COST

Compound		A	B
		-Pounds-	
PVC		100	100
DOP		50	40
Fast-Solvating Plasticizer		–	10
Stabilizer/Lubricant		3	3
Total Pounds		153	153
	Sp. Gr.	Pound Volumes	
PVC	1.4	71.43	71.43
DOP	0.986	50.71	40.57
Fast-Solvating Plasticizer	1.12	–	8.93
Stabilizer/Lubricant	1.00	3.00	3.00
Total Pound Volumes		125.14	123.93
	¢/Pound	¢	
PVC	42	4200	4200
DOP	47	2350	1880
Fast-Solvating Plasticizer	50	–	500
Stabilizer/Lubricant	100	300	300
Total¢		6850	6880
Sp. Gr.		1.223	1.235
¢/Pound (Total¢/Total Pounds)		44.77	44.97
¢/Pound Volume (Total¢/Total Pound Volumes)		54.74	55.52
$/Ft^3 62.3(¢/Pound Volume/100)		34.10	34.59
$/100 yard roll ($/Ft^3 x 1.35)		46.04	46.70

Added cost/100 yard roll = 46.70 − 46.04 = $0.66.

mixing, fluxing, straining, and calendering energy consumption.
Since the one hundred yard roll of 54 inch width 12 mil sheeting
equals 1.35 ft.^{3} and weighs 1.35 x 62.3 x 1.223 = 103 lbs (in the
case of Compound A), energy cost would be $2.06. Thus, energy
usage would have to be cut by 32 percent just to offset the 66¢
increase in raw materials cost. Energy savings of this magnitude
seem unlikely.

In some cases, taking the opposite approach, using slower
solvating, higher molecular weight and less efficient phthalates,
such as DINP and DIDP, has been shown to increase plant output in
certain calendering and extrusion operations. Table II compares
results of a wire insulation extrusion study with DOP vs. DINP.
It was found that at comparable temperatures and screw speeds,
output of the DINP formulation was 17 percent higher on a volume
basis than that of the DOP formulation. In this example 58 phr
(parts per hundred of resin) of DINP was compared with 55 phr of
DOP (equal efficiency levels).

The importance of comparing plasticizers at equal efficiency
levels (equal hardness or 100% modulus) rather than equal phr
cannot be overemphasized. This is shown in Table III for a typical plastisol compound. It can be seen that use of a less efficient plasticizer, such as DIDP, even though it costs one half
cent per pound more than DOP, results in a pound volume cost
savings of 0.7¢.

For those who calculate costs on a weight basis the higher
cost per pound of the DIDP compound can be more than compensated
for by addition of extra filler. For example, as shown in Table
III adding an extra 7 phr of $CaCO_3$ and an extra phr of DIDP
equates specific gravity to that of the DOP compound. This
results in a savings of 1.17 cents per pound or 1.51 cents per
pound volume.

The point of the foregoing was to illustrate the importance
of considering all aspects of formulation changes – energy
savings may or may not be real, and the extra compound cost, plus
possible adverse affects on overall performance must also be
considered.

5. <u>Other Factors Influencing the Use of Monohydric Alcohol
Based PVC Plasticizers in the Eighties</u>. Increasing attention to
the cost of fuming losses – especially in calendering and spread
coatings and the increasing restrictions on stack emissions and
the in-plant environment will tend to continue the trend to less
volatile plasticizers – i.e., higher molecular weight phthalates
than DOP. Figure 1 shows comparative volatilities of various GP
plasticizers(<u>23</u>). It is obvious that DOP volatility during the
fusion process is much greater than that of higher dialkyl
phthalates. Aside from improving the environment, there is an
opportunity then to cut economic losses through selection of
lower volatility plasticizers.

The use of hydrocarbon plasticizers as partial replacements
for monohydric alcohol phthalates may tend to decrease as a result

TABLE II

<u>WIRE INSULATION EXTRUSION STUDY</u>

(0.8 mm Wall Thickness)

Plasticizer (phr)	DOP(50)	DINP(58)
<u>Extruder Temperature, °C</u>		
Rear Zone	174	171
Middle Zone	182	183
Front Zone	187	188
Die	190	191
Screw Speed, RPM	118	118
Volume Output, % of DOP	100	117

<u>Formulation</u>

PVC	100
Plasticizer	as shown
Lead Stabilizer	5
$CaCO_3$	10
Calcined Clay	10
Stearic Acid	0.125

TABLE III

ECONOMIC COMPARISON OF DOP VS. DIDP IN PLASTISOL COMPOUNDS

Compound		C	D	E
			−Pounds−	
PVC		100	100	100
DOP		60	−	−
DIDP		−	66	67
Stabilizer		3	3	3
$CaCO_3$		20	20	27
Total Pounds		183	189	197
	Sp. Gr.	Pound Volumes = Pounds/Sp. Gr.		
PVC	1.4	71.43	71.43	71.43
DOP	0.986	60.85	−	−
DIDP	0.968	−	68.18	69.21
Stabilizer	1.00	3.00	3.00	3.00
$CaCO_3$	2.71	7.38	7.38	9.96
Total Pound Volumes		142.66	149.99	153.60
	¢/Pound	¢ = ¢/Pound x Pounds		
PVC	50	5000	5000	5000
DOP	47	2820	−	−
DIDP	47.5	−	3135	3183
Stabilizer	100	300	300	300
$CaCO_3$	5	100	100	135
Total¢		8220	8535	8618
Sp. Gr.		1.283	1.260	1.283
¢/Pound (Total¢/ Total Pounds)		44.92	45.16	43.75
¢/Pound Volume (Total¢/Total Pound Volumes)		57.62	56.90	56.11
¢/Pound Volume Savings (vs. DOP)		−	0.72	1.51

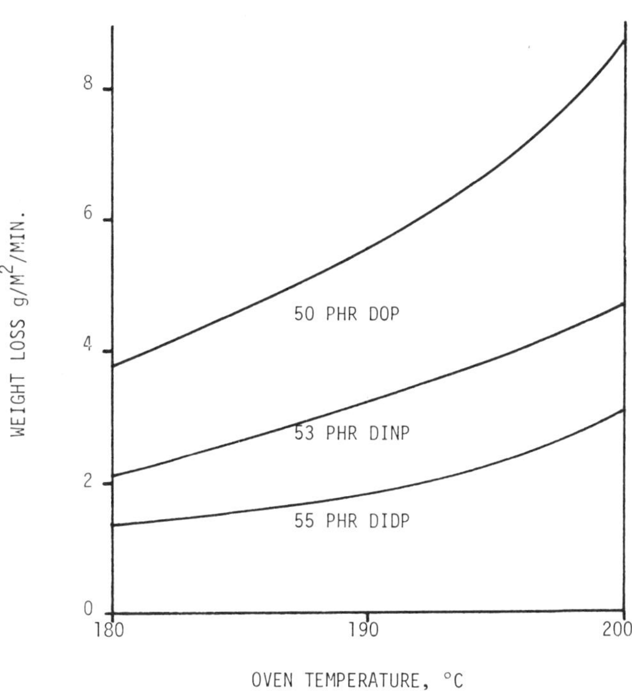

Figure 1.   Weight loss from spread coating (Werner–Mathis oven; coating thickness: 0.25 mm; dwell time: 2 min)

again of more restrictive air pollution controls. In addition,
there is increasing concern about the toxicity of certain highly
aromatic types used, particularly in low cost plastisol formulations. In those cases where hydrocarbon type plasticizers are
used, the use of higher molecular weight phthalates, such as DIDP,
will overcome to some extent the overall high volatility.

The use of trimellitates is expected to regain ground lost
due to TMA shortages in the late seventies. Polymerics will continue to be used in those areas where high solvent extraction
resistance and migration are major concerns. Phosphates are
expected to continue their growth in flame retardant applications. Also, new capacity for dioctyl terephthalate (DOTP) has
been announced(25).

Finally, a recently published draft report of a National
Toxicology Program study found DOP (or DEHP as it is referred to
in toxicology studies) to be carcinogenic to certain strains of
mice and rats when fed at very high levels in the diet for the
lifetime of the test animals. It may be a number of years before
the mechanism of the induced carcinogenic effects can be understood, and a realistic assessment of any risk to man can be made.
The final version of the draft report is expected to issue in
February, 1981.

Conclusions

It appears that monohydric alcohol phthalates will continue
to dominate the vinyl plasticizer market in the eighties. But,
the growth of specific types of alcohols is continuing to
change - a trend that started in the fifties and sixties, became
a landslide in the seventies, and is expected to continue in the
eighties. Increased attention to raw material and energy costs
and production rates, will be required for the vinyl fabricator
to remain competitive. In addition, the overall impact of formulation changes, aside from economics, must be more carefully
weighed, including affect on physical properties, fuming losses,
and the environment.

Literature Cited

1. Semon, W. L. U.S. Pat. 1,929,453, Oct. 10, 1933.
2. Kyrides, L. P. U.S. Pat. 1,923,938, Aug. 22, 1933.
3. Corley, H. M. "Successful Commercial Chemical Development"; John Wiley & Sons, Inc.: New York, 1954; p. 341.
4. Anon. Mod. Plastics, 1945, 23(3), 169.
5. Schatzel, A. W. Private communication, Oct. 16, 1980.
6. Powell, G. M. and Quarles, R. W. Official Digest, Fed. Paint Varnish Prod. Clubs, 1946, 263, 696–702.
7. Hatch, L. F. "Higher Oxo Alcohols"; John Wiley and Sons, Inc.: New York, 1957; pp. 1–2.
8. Darby, J. R.; and Sears, J. K. "Plasticizers" in Encyclopedia of Chemical Technology; John Wiley & Sons, Inc.: New York, 1968; Vol. 15, p. 723.
9. Dimler, W. A.; Mountain, J. A. and Overberger, W. F. SPE Technical Papers, 1960, Vol. 6; p. 57.
10. Quackenbos, H. M., Jr. Ind. Eng. Chem., 1954, 46, 1335–1344.
11. Towle, P. H.; Baldwin, R. H. and Meyer, D. H. "Phthalic Acids" in Encyclopedia of Chemical Technology; John Wiley & Sons, Inc.: New York, 1968; Vol. 15, p. 458.
12. Brit. Pat. 851,753 Oct. 19, 1960.
13. U.S. Tariff Commission Data.
14. Skeist, I. Chem. Week, 1955, 76(16), 40–57.
15. Fusco, S. J.; Maggart, R. C. and Overberger, W. F. "Linear Oxo Phthalate Plasticizers" in "Plasticization and Plasticizer Processes", American Chemical Society: Washington, D.C., 1965, p. 62.
16. Anon. Chem. Marketing Reporter, 1980, 218(13), 3.
17. Beeler, A. D. Plastics Engineering, 1976 32(7), 40–41.
18. Anon. Chem. Marketing Reporter, Sept. 3, 1979, p. 3, 23.
19. Anon. C&EN, 1980, 58(44), 17.
20. Anon. Kagaku Kogyo Nippo (Japan), July 16, 1980.
21. Clementi, A.; Oriani, G.; Ancillotti, F., and Pecci, G. Hydrocarbon Processing, 1979, 59(12) 109–113.
22. Eighmy, G. W. Private communication, Oct. 16, 1980.
23. Poppe, A. C. Kunststoffe, 1980, 70(1), 38–40.
24. Anon. Plastics Technology, 1980, 26(10), 104.
25. Anon. Chem. Week, 1980, 127(16), 14.

RECEIVED December 29, 1980.

Monohydric Alcohols in the Flavor and Fragrance Industry

J. DORSKY

Givaudan Corporation, Research Department, Clifton, NJ 07014

Todays' perfumers and flavorists have a palette of about 5,000 materials at their disposal to formulate the pleasant odors and tasty flavors enjoyed and even demanded in consumer products. Some materials are of natural origin, others are purely synthetic and some are available from both sources, natural and synthetic. Almost all kinds of organic functionalities are represented in the broad palette of the flavorist and perfumer - alcohols, aldehydes, ketones, esters, hydrocarbons, olefins, amines, phenols, heterocyclics, etc. Alcohols are particularly important because they are prominent among the relatively inexpensive and readily available materials which make up the bulk of flavors and especially fragrances.

Flavors vs. Fragrances

Flavors and fragrances are sensory stimuli. Of the two, flavors are more complex because they act on the olfactory bulb via their volatile components and on the taste buds which are stimulated by both volatile and non-volatile components. The overall response to a flavor is a synthesis of the effects of both types of components. The response to fragrances, on the other hand, results only from the action of volatile components. Because flavors and fragrances function via a common mechanism, many volatile materials are used for both purposes. This is nicely illustrated by the perfumers' vocabulary for fragrance materials. A collection of some 160 words published by a famous perfumer, Ernest Shiftan (1) included 75 words usually associated with flavors such as almond, bacon, coconut, honey, lime, raspberry, spicy and vanilla.

Variety and Diversity of Monohydric Alcohols

Givaudan Corporation, probably the largest supplier of flavor and fragrance chemicals, offers about 50 monohydric alcohols. This is by no means the whole picture. Other suppliers offer additional alcohols and some alcohols are used in-house as intermediates,

RHODINOL	50%
GERANIOL COEUR	5
CITRONELLOL COEUR	10
PHENYLETHYL ALCOHOL COEUR	10
NEROL COEUR	5
GERANYL ACETATE	2
ALDEHYDE C-8 10%	4
ALDEHYDE C-9 10%	4
BENZOPHENONE	3.5
ROSE OXIDE 1%	1
ROSALVA	0.5
ESSENCE OF STYRAX	3
GUAIACWOOD OIL (GUAIOL)	2
TOTAL	100%

Figure 1.   Rose base

ALDEHYDES	45.0%
ALCOHOLS[*]	48.0
ESTERS	4.0
ACETALS	0.4
PHENOL ETHERS	0.7
KETONES	0.3
NITROGEN COMPOUNDS	0.5
RESINS	0.1
OTHERS	1.0

[*] HYDROXYCITRONELLAL, TERPINEOL, LINALOOL,
PHENYLETHYL ALCOHOL, ANISIC ALCOHOL,
DIMETHYLBENZYL CARBINOL, CINNAMIC ALCOHOL

Figure 2.   Lilac perfume

or are only sold in fragrance mixtures. The number of monohydric
alcohols used by the industry is no doubt greater than 100.

The importance of monohydric alcohols in fragrances is clearly shown in the formula for a rose base in Figure 1 which contains
82.5% monohydric alcohols. The chemistry of some of these alcohols is covered later.

A typical lilac perfume in Figure 2 contains 48% alcohols,
all monohydric.

Nature has been very generous in distributing fragrant alcohols among the flowers. Figure 3 lists the number of different
alcohols found in four important floral extracts used in perfumery:
jasmin, hyacinth, ylang-ylang and tuberose. The number ranges
from 31 to 56 and terpenoid alcohols are the most abundant. Concentrations of individual alcohols in essential oils vary widely,
from traces to as much as 80%, in the case of santalol in sandalwood oil.

Nor has nature neglected to bless fruits with alcohols which
add their special touch to the flavors of fresh fruit. A list of
alcohols found in apple and banana extracts is shown in Figure 4.
These examples suffice to show how widely distributed alcohols are
in natural flavors and fragrances.

Chemical Types

Before the 19th century, perfumers had at their disposal only
natural products of plant and animal origin. Today, perfumers
work with about 5,000 materials, most of which are produced synthetically. Among the approximately 100 monohydric alcohols used, all
chemical types are represented. Several examples of aliphatic alcohols are shown in Figure 5. Nonyl alcohol and 3-octanol are
typical fatty alcohols. These alcohols are often found in natural
products but are seldom used because of their weak odor.

In contrast, olefinic alcohols have strong odors and are widely used in perfumes, though at relatively low concentrations. Leaf
alcohol has a very pleasant leafy, green odor. It is found in
many flowers and fruits and is the most important of the olefinic
alcohols. Commercial leaf alcohol contains at least 90% of the
cis isomer.

Typical aromatic alcohols are shown in Figure 6. Phenylethyl
alcohol is the most important member of this family. It is the
main constituent of French rose and is also present in Otto of rose.
Phenylethyl alcohol has a heavy, sweet odor reminiscent of rose
petals. It blends well with other floral odors and is widely used
in many floral fragrances. Phenylethyl alcohol is one of the chemical pillars of perfumery because, in addition to its fine odor
qualities, it is relatively inexpensive and readily available.

Cinnamic alcohol, anisic alcohol and dimethylphenyl ethyl carbinol are other members of the aromatic alcohol family.

	ALIPHATIC	AROMATIC	TERPENOID	TOTAL
JASMIN	10	4	17	31
HYACINTH	11	12	33	56
YLANG YLANG	10	7	31	48
TUBEROSE	9	4	21	34

Figure 3.   Number of alcohols found in natural floral extracts

	APPLE	BANANA
1-BUTANOL	X	X
2-BUTANOL		X
ETHANOL	X	X
GERANIOL	X	
1-HEXANOL	X	X
TRANS-2-HEXEN-1-OL	X	
CIS-3-HEXEN-1-OL	X	
METHANOL	X	
2-METHYLBUTAN-1-OL	X	
3-METHYLBUTAN-1-OL	X	X
2-METHYLPROPAN-2-OL	X	X
1-PENTANOL	X	
2-PENTANOL		X
1-PROPANOL	X	X
2-PROPANOL	X	

Figure 4.   Alcohols found in fruit extracts

FATTY ALCOHOLS

ODOR TYPE

$CH_3(CH_2)_7CH_2OH$

FATTY – ORANGE

NONYL ALCOHOL

OH
$CH_3CH_2CH(CH_2)_4CH_3$

PEPPERMINT-SAGE

3-OCTANOL

OLEFINIC ALCOHOLS

GREEN LEAF

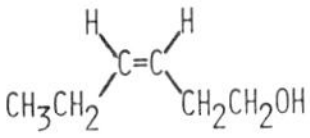

LEAF ALCOHOL

cis-3-HEXENOL

OH
$CH_3(CH_2)_4CH-CH=CH_2$

MUSHROOM

VERY STRONG

1-OCTEN-3-OL

Figure 5.

ODOR TYPE

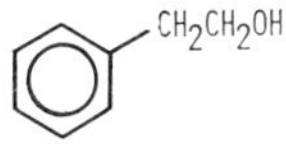

CH_2CH_2OH

ROSE-HONEY

PHENYL ETHYL ALCOHOL

$CH=CH\ CH_2OH$

BALSAMIC-FLORAL

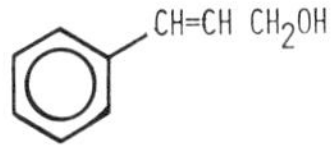

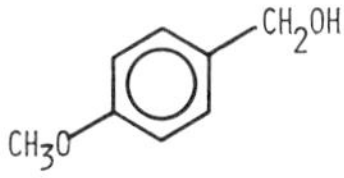

CINNAMIC ALCOHOL

CH_2OH

FLORAL
BALSAMIC

CH_3O

ANISYL ALCOHOL

OH
$CH_2CH_2C-CH_3$
CH_3

LILY
HYACINTH

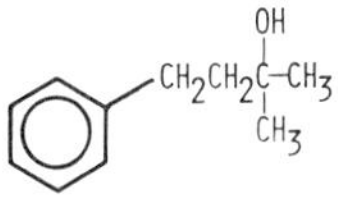

DIMETHYLPHENYLETHYL
CARBINOL

Figure 6.   Aromatic alcohols

Terpenoid alcohols appeared early in the history of synthetic perfumery because several were readily available from inexpensive essential oils. Alpha-terpineol, citronellol and linalool shown in Figure 7 are important constituents of pine stump oil, citronella oil and rosewood oil, respectively. The fourth material listed, hydroxycitronellal, is a hydroxy aldehyde which perhaps has a questionable place in this discussion. It is included because it is one of the most important fragrance chemicals used today. "Hydroxy" is almost a perfume unto itself. Its soft flowery, linden blossom odor blends very well in many floral perfumes.

Sesquiterpenic alcohols used in perfumery are mainly of natural origin. Three alcohols of this type are shown in Figure 8. Cedrol is the main alcohol constituent of cedarwood oil. Alphasantalol constitutes about 80% of sandalwood oil and about 30% patchouli alcohol is found in patchouli oil, a very popular woodyearthy fragrance material.

Ethyl Alcohol

Perfumes, colognes and toilet waters are solutions of perfume oils in specially denatured alcohols, (ethyl alcohol with various denaturants which make it unsuitable for human consumption). In addition, ethanol is used as a solvent for flavors, as an extraction solvent for many natural products and as a reagent for the production of many ethyl esters. Summing up all these uses, the quantity of ethanol used in the fragrance and flavor industry tops all other alcohols by far.

Perfumes contain 10-25% oil in SDA 39C. The denaturant is diethyl phthalate - 1:100. Colognes and toilet waters contain 2-6% oil in SDA 40 as shown in Figure 9. The denaturant is brucine sulfate - 3 oz. per 100 gallons. Ethanol used as a solvent for flavors is the 95% grade and is fully taxed since it is potable. All ethanol used for fragrances and flavors must get by the close scrutiny of the Quality Control perfumer or flavorist. The road can be rough.

Hydroxycitronellal

This is probably the single most important material used by the fragrance industry. Several million pounds are used annually, mainly in soaps and detergents. The principal method of manufacture shown in Figure 10 is by hydration of citronellal via the bisulfite addition product (2). The aldehyde moiety must be protected before hydration. A second manufacturing process starts with citronellol which is hydrated under acid conditions. The primary alcohol end of the molecule is then dehydrogenated catalytically or by oxidation to the aldehyde.

ODOR TYPE

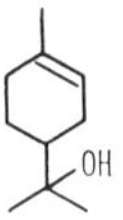

LILAC

ALPHA-TERPINEOL

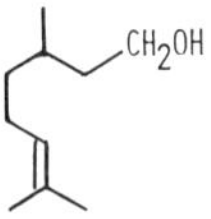

FRESH - ROSEY

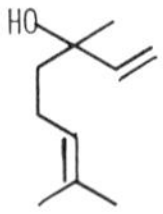

CITRONELLOL

FLORAL - WOODY

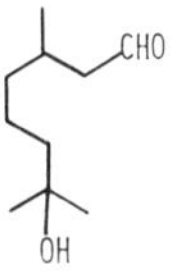

LINALOOL

LILY-OF-THE-VALLEY

HYDROXYCITRONELLAL

Figure 7.   *"Terpenoid" alcohols*

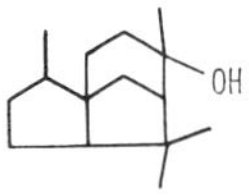

CEDROL

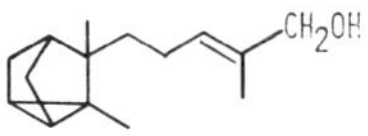

ALPHA-SANTALOL

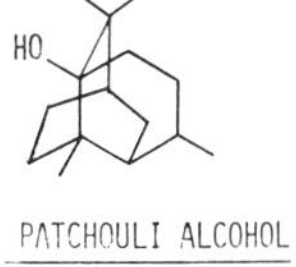

PATCHOULI ALCOHOL

Figure 8.   *Sesquiterpenic alcohols*

PERFUME

10-25% OIL IN 95% 39C ALCOHOL

COLOGNE/TOILET WATER

2-6% OIL IN 60-80% 40 ALCOHOL

Figure 9. Ethyl alcohol

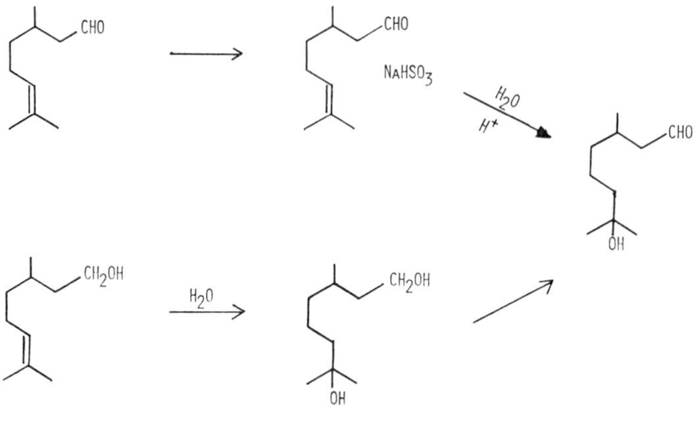

Figure 10. Hydroxycitronellal

Menthol

Peppermint flavor and "cool" cigarettes owe their distinctive flavor character to l-menthol, a secondary alcohol and another multimillion-pound product of the industry. Pure l-menthol is used mainly for flavoring cigarettes and much of this material is isolated from the _Mentha arvensis_ plant. Substantial quantities are also produced synthetically from meta-cresol as shown in Figure 11. Isopropylation yields a mixture of isomers from which thymol is isolated. Hydrogenation of thymol gives a mixture of menthol isomers. Racemic menthol is recovered from the mixture by careful fractionation of menthol as an ester. l-Menthol is the desired isomer because it is responsible for the cooling effect. Several practical methods have been described for separating the l-isomer. One patented by Haarmann and Reimer is especially intriguing (3). A supersaturated solution of racemic menthol esters is seeded with the l-ester which then selectively crystallizes.

Geraniol/Citronellol

These "rose" alcohols were formerly obtained from citronella oil as a 2:1 mixture of geraniol/citronellol. High purity citronellol was obtained by hydrogenation of the mixture. Soap perfumery was highly dependent on cheap natural geraniol/citronellol. Since the 1960's, synthetic geraniol has been a controlling factor in the market for these alcohols. The synthetic route is shown in Figure 12. Myrcene, produced by pyrolysis of beta-pinene, is converted to a mixture of geraniol and nerol, its _cis_ isomer, first by reaction with HCl, then with sodium acetate and finally with sodium hydroxide to saponify the esters.

Commercialization of this chemistry by Glidden (4) was the first case where synthesis freed the industry from total dependence on a natural source of a major product.

Linalool

In the mid-1950's, Hoffmann La-Roche offered synthetic linalool and its esters to the fragrance industry (5). This was another revolutionary step in the replacement of natural products with synthetics. Linalool was an intermediate in Roche's manufacture of synthetic isophytol for Vitamin E. For several years perfumers were reluctant to use chemically pure synthetic linalool and its esters in place of materials isolated from essential oils. Finally, economics prevailed and today the amount of synthetic linalool and its esters used far exceeds the traditional "linalool rich" essential oils.

Worldwide, in excess of five million pounds of linalool and its esters are manufactured for fragrances by five different processes. Three of them involve methylheptenone as an intermediate. The other two are based on turpentine. The three routes to methylheptenone are outlined in Figure 13.

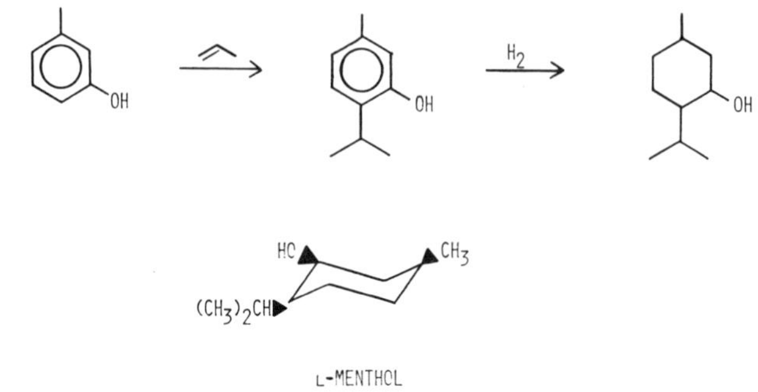

Figure 11.   Menthol

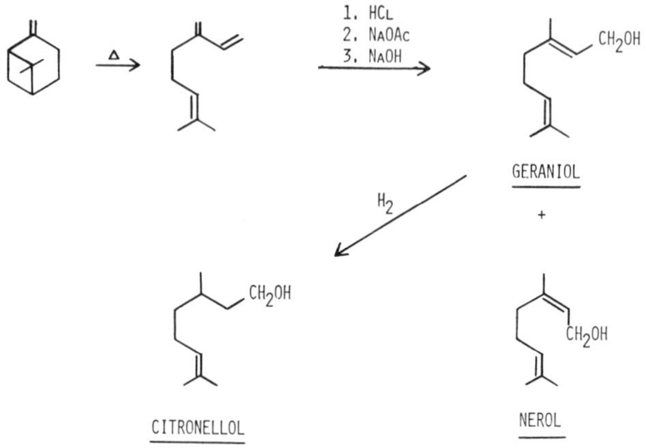

Figure 12.   Geraniol/citronellol

In the first route, methylbutenol is made from acetone and acetylene followed by hydrogenation. Reaction with methyl isopropenyl ether yields methylheptenone (6). The second route involves the reaction of isobutylene, formaldehyde and acetone (7). Methyl vinyl ketone is an intermediate. Finally, methylheptenone is made by alkylation of acetone with prenyl chloride which is derived from isoprene (8). The initial product is the terminal olefin which is isomerized to the desired isopropylidene compound.

As shown in Figure 14, linalool is obtained by ethynylation of methylheptenone, followed by hydrogenation.

Sandalore

Some of the chemistry developed by the industry more recently, to produce new monohydric alcohols, is just as interesting as the linalool chemistry. Sandalore, a recent new Givaudan chemical with a persistent, sandalwood odor is made according to the scheme in Figure 15 (9). Alpha-pinene, the starting material, is converted to the epoxide which is catalytically rearranged to campholenic aldehyde. Aldol condensation with methyl ethyl ketone followed by hydrogenation yields Sandalore®.

Phenylethyl Alcohol

This important fragrance material probably was introduced in commercial perfumery during the first decade of the twentieth century with the discovery of the Bouveault-Blanc reduction of esters by sodium and an alcohol (10). This is the first of several methods of preparation shown in Figure 16. Large quantities of phenylethyl alcohol were made by sodium reduction of butyl phenylacetate in normal butanol. The basic raw materials were readily available at low cost from benzyl chloride, sodium cyanide and fermentation butanol.

This process was superceded by the Friedel-Crafts reaction of benzene and ethylene oxide which is now the most important commercial process (11). Cheap ethylene oxide brought about this shift. For a short time, commercial quantities of phenylethyl alcohol were made by the Grignard reaction of phenyl magnesium chloride on ethylene oxide but this process could not compete with the Friedel-Crafts process.

More recently, low cost styrene oxide has been considered as a starting material. Catalytic hydrogenation under controlled conditions yields high quality phenylethyl alcohol (12). Unfortunately, low cost styrene oxide never materialized in this country; consequently, little phenylethyl alcohol has been manufactured by this route.

The four methods of producing phenylethyl alcohol depicted in Figure 16 illustrate the dependence of the fragrance industry on the organic chemical industry. Most large volume products used in

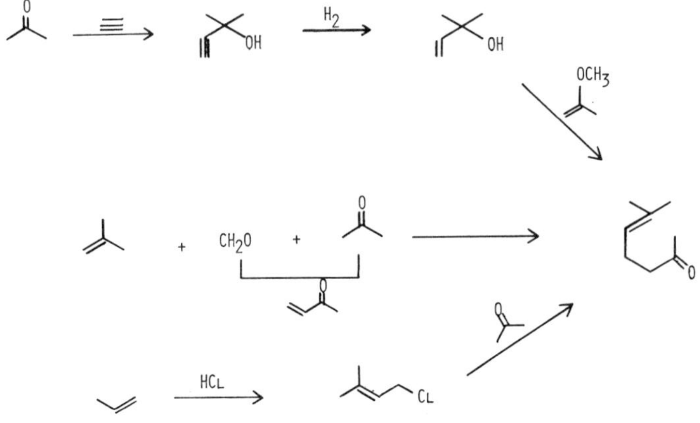

Figure 13.   Methylheptenone

Figure 14.   Linalool

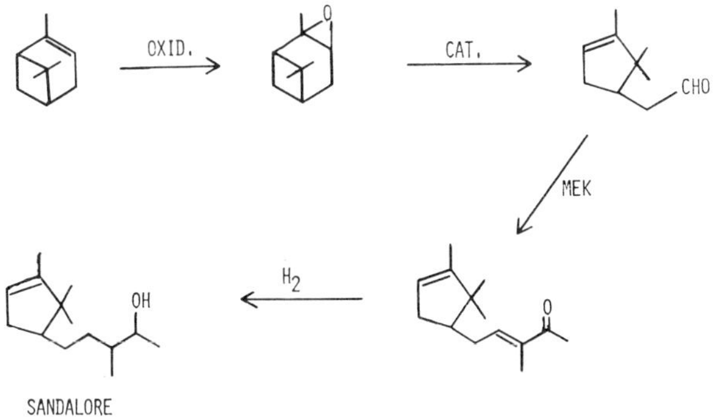

Figure 15. *Sandalore*

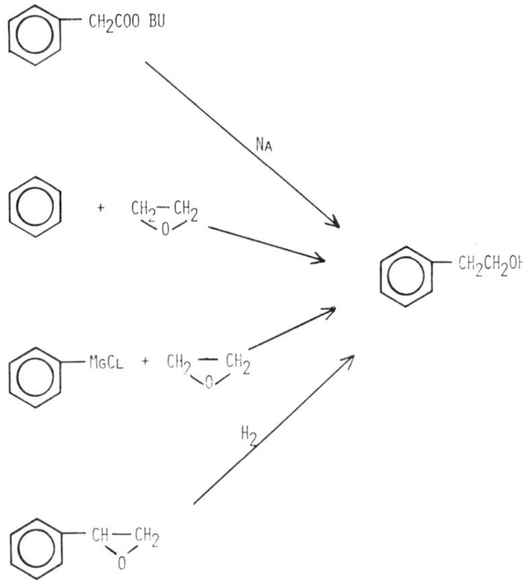

Figure 16. *Phenylethyl alcohol*

perfumes are commodity items that must be made by the least expensive process. Very often, raw materials cost determines which process will prevail.

This overview of monohydric alcohols covered the major products used by the industry. The one exception was Sandalore which was introduced only recently. It was included as an example of the continuing search for new fragrance materials to replace natural products, in this case, expensive and scarce sandalwood oil. It is hoped that a similar review in five years would include Sandalore or another major new, synthetic fragrance alcohol.

Abstract

Monohydric alcohols are found in great variety in natural flavor and fragrance materials. Many chemical types are represented: aliphatic, alicyclic, polycyclic, heterocyclic, terpenoid, saturated, unsaturated, etc. In flavor and odor they cover practically the entire useful spectrum. Some are present in low concentrations and others are the principal components of the natural products. Flavorists and perfumers first formulated sophisticated flavors and fragrances using naturally occurring alcohols. As the art and science progressed, chemists devised synthetic routes to many of the important natural alcohols and produced them cheaper and in higher quality. Commercial synthesis of some natural alcohols, such as, geraniol, citronellol, menthol and linalool has had a very significant impact on the economics of the flavor and fragrance industry. The preparation and use of the more important monohydric alcohols is reviewed.

Literature Cited

1. Shiftan, E., _Encyclopedia of Chemical Tech._, 2nd Ed., John Wiley & Sons, New York, 1967, 14, 744.
2. Meuly, W.C., U.S. Pat. 2,235,840, March 25, 1941.
3. Haarman & Reimer, GMBH, Brit. Pat. 1,369,714, October 9, 1974.
4. Webb, R.L., U.S. Pat. 3,031,442, April 24, 1962.
5. Offner, A., _Soap, Perfumes, Cosmetics_, February 1965, 38, 125.
6. Saucy, G. and Marbet, R., _Helv. Chim. Acta._, 1967, 50, 2091.
7. Pommer, H., Müller, H. and Overwien, H., Ger. Pat. 1,259,876, May 11, 1966. Ger. Pat. 1,268,135, July 16, 1966.
8. Meuly, W.C., _Riechstoffe, Aroman, Körperpflegemittel_, 1972, 6, 191.
9. Naipawer, R.E. and Easter, W., U.S. Pat. 4,052,341, October 4, 1977.
10. Bouveault, L. and Blanc, G., _Compt. Rend._, 1903, 136, 1676, _Bull. Soc. Chim. France_, 1904, 31 (3) 666.
11. Carpenter, S., U.S. Pat. 2,013,710, 1935.
12. Wood, T.F., U.S. Pat. 3,579,593, May 18, 1971.

RECEIVED December 29, 1980.

INDEX

A

Acetals	7
Acetate, *n*-butyl	82
Acetate, isobutyl	83
Acetic acid, conversion of methanol to	25
Acetylenes, alcohol addition to	4–5
Acids	
acid-catalyzed esterification of organic	6
hydrogenation of fatty	89
reaction of alcohols with organic	5–7
Acrylate, *n*-butyl	81
Acrylate, 2-ethylhexyl	84
Acrylic–melamine resin coating systems	81
Acrylic resin coating systems, thermoset	81
Alcohol(s)	
with an acyl halide, reaction of	6
addition to acetylenes	4–5
with aldehydes and ketones, reactions of	7
aliphatic	199
to alkyl halides, converting	9
to amines, conversion of	9
analytical determination of the hydroxyl group in	14
aromatic	199, 201*f*
capryl (octanol-2)	182
with carboxylic acid, reaction of	165
changing value of olefin feedstocks for	187
by chromium (VI) species, oxidation of	12
conversion of cellulose to fermentable sugar for conversion to	52
conversion of olefins to	159–163
dehydration of	10–12
to detergents, ethoxylation of	159
detergent-range	87, 163
displacement of the hydroxyl group in	9–10
EPAL	99
esterification	5–7
ether sulfate(s)	101, 107–111
heavy-duty laundry powders	107
light-duty dishwashing liquids	107
shampoos	109

Alcohol(s) (*continued*)

ethoxylates	101
primary (PAE)	117
secondary (SAE)	113–156
biodegradability of	124
continuous-flow activated sludge test for	127, 128*f*
river die-away test for	124–127
calcium ion sequestering ability	138, 142*f*
characteristics of	133*t*
cloud point of	119, 121*f*
derivatives of	146–152
detergency in dishwashing	133, 138, 139*f*
detergency, heavy-duty laundry application	133, 134*f*, 135*t*, 136*f*
ethoxysulfates from	149
foam properties of	119, 123*f*
good rinsability of	141
industrial applications of	141–146
low-temperature engine oil viscosity classification	175*t*
manufacturing process of	115
performance in household detergents	131–140
physical and surface active properties of	117–124
pour point	117, 120*f*
propylene oxide adducts of	146
safety and environmental assessment of	124–131
sulfosuccinate half esters of	149–152
toxicity to fish	127–131
uses in the pulp and paper industry	145
uses in the textile industry	141–145
in various lubricant base oils, miscibility of $C_{12\text{-}14}$	144*t*
viscosity	119, 120*f*, 121*f*
classifications for motor oils	170, 172*t*
and clear point behavior in liquid formulation	138, 140*f*
wetting power of	119, 122*f*
ethyl	202, 204*f*
fatty	87, 201*f*
synthetic	88

Alcohol(s) (*continued*)
  in fruit(s) 199
    extracts 200*f*
  fuels
    from biomass, government
        incentives (tax credits)
        to produce 56–58
    process economics, grain 59, 61
    from sources other than
        petroleum, production of 56
  in gasohol, impact of incentives
    cost of 62*t*
  hexyl 184
  in household surfactants, appli-
    cations of higher 101–112
  with isocyanates, reaction of 8
  isodecyl 182
  isononyl 185
  isooctyl 182
  leaf 199
  manufacture, fermentation 49–51
  manufacture, synthetic 47–49
  monohydric
    acidity and basicity of 1–2
    chemistry of saturated aliphatic .. 1–15
    ester plasticizers for (PVC) ... 181–194
    in the flavor and fragrance
        industry 197–210
    phthalates 181–194
    secondary 182
    synthetic lubricant basestocks
        from 165–179
    variety and diversity of 197
  nonionic(s) 101, 103–107
    heavy-duty laundry liquids ... 105–107
    heavy-duty laundry powders .. 103–105
  in natural floral extracts,
    number of 200*f*
  nucleophilic reactions of 2–9
  olefinic 199, 201*f*
  with organic acids, reaction of 5–7
  oxidation of 12–14
  oxo 88
    C_4 79
    manufacture, higher linear ... 159–163
  phenylethyl 199, 207, 209*f*
  plasticizer 87
    range 163
  process, ALFOL 92*f*
    Conoco 93–95
  process, ethyl-modified linear 95–99
  production of natural 88–90
  production, use of coal in 63
  190-proof 53
  purification of crude 161
  rose 205
  secondary 88
    manufacturing process of 115
  sesquiterpenic 202, 203*f*

Alcohol(s) (*continued*)
  straight-chain 184
    by the ethylene chain growth
        process, manufacture of
        higher 87–99
  sulfate(s) 111
    shampoos 111
  terpenoid 202, 203*f*
  tertiary butyl (TBA) 65
  Ziegler 88
    production of 90–93
Alcoholysis of anhydrides 7
Alcoholysis of methacrylamide with
    n-butanol 81
Aldehydes and ketones, reactions of
    alcohols with 7
ALFOL alcohol process 92*f*
  Conoco 93–95
Aliphatic alcohols 199
  monohydric, chemistry of saturated 1–15
Alkyl
  dimethyl amine oxides 102
  glyceryl ether sulfonates 102, 112
  halide, S_N2 displacement on 3
  sulfates 102
Alkylbenzene sulfonate(s) (LAS) ... 107, 109
Alkoxide ion, nucleophilic reactions of   3–4
Alkoxides, hydrolysis of aluminum ... 93
Alkoxylation 8
Alumina, CATAPAL 94
Aluminum alkoxides, hydrolysis of ... 93
Amine(s)
  C_4 82
  conversion of alcohols to 9
  oxides 102, 112
    alkyl dimethyl 102
Amino resin(s) 23
  systems 81
Ammonolysis 10
Anhydrides, alcoholysis of 7
API passenger car engine oil
  classification, current 176*t*
Aromatic alcohols 199, 201*f*
Arylnitroalkenes 4
Automotive
  fuel/engine systems, compatibility
      of ethanol–gasoline blends
      with 68
  crankcase lubricants, partial
      synthetic 170
  diester vs. petroleum
      basestocks of 179
  upholstery, linear phthalates in 187
Azelaic acid, 2-ethylhexanol diester of   174

B

Bacteria 25
BBP (butyl benzyl phthalate) 82, 183
Beer 49

Benzene and ethylene, Friedel–Crafts,
reaction of 207
Biomass
-derived ethanol 55
1980–1995, potential demand
for 59, 60*f*
production, net energy gain in 63
energy in Brazil 30
Energy Security Corporation
programs for 57
feedstocks potentially available
for ethanol fuel production 62*t*
government incentives (tax credits)
to produce alcohol fuels from..56–58
investment costs of methanol
production from coal, lignite and 35
liquid fuels from 30
Boiler fuel 63
BOP (butyl octyl phthalate) 183
Bouveanult–Blanc reduction of esters 207
Brazil
biomass energy in 30
methanol from wood in 29–44
primary energy consumption in
1977 .... 29*t*
Bucherer reaction 9
n-Butanol 183
alcoholysis of methacrylamide with 81
applications of 79–82
consumption pattern in the U.S..... 80*t*
by the rhodium oxo process,
manufacture of 71–84
n-Butenes 187
Butyl
alcohol, tertiary (TBA) 65
benzyl phthalate (BBP) 82, 183
ether(s)
CARBITOL diethylene glycol
mono-*n*-butyl 81
CELLOSOLVE (ethylene glycol
mono-*n*-butyl 81
methyl tertiary (MTBE) 65
glycol 81
octyl phthalate (BOP) 183
n-Butyl
acetate 82
acrylate 81
methacrylate 81
Butylamines 82
Butylenes 187
Butyraldehyde hydrogenation and
refining 77
n-Butyraldehyde 71, 183, 185

C

CO and H_2 to methanol, conversion of 15
Calcium ion sequestering ability,
SAE 138, 142*f*

Capryl alcohol (octanol-2) 182
Car engine oil(s) classification(s),
passenger 174
current API 176*t*
CARBITOL, butyl, diethylene glycol
mono-*n*-butyl ether 81
Carbon source for single-cell protein
production, methanol as 25
Carbon source for water
denitrification, methanol as 26
Carbonyl catalyst, cobalt 73
Carboxylic acid, reaction of alcohol
with 165
Carcinogen, ethanol as a suspect 53
Catalyst(s)
cobalt
carbonyl 73
ligand 73
system, oxo 160
oxo 73
system 160
CATAPAL alumina 94
CELLOSOLVE, butyl, ethylene
glycol mono-*n*-butyl ether 81
Cellulose 50
ethanol production via hydrolysis of 50
as a feedstock for the production
of ethanol 51
to fermentable sugar for conversion
to alcohol, conversion
of 52
to sugars, degradation of 50
Cellulosic plasticizers 181
CESP's pilot plant, methanol
synthesis— 41*f*
Chain growth process, manufacture
of higher straight-chain alcohols
by the ethylene 87–99
Charcoal gasification 39–43
Chlorohydrins into epoxides,
transformation of 3
Chromium (VI) species, oxidation
of alcohols by 12
Citronellol 205, 206*f*
Clostridium thermocellum 51
Clostridium thermosaccharolyticum .. 51
Coal
in alcohol production, use of 63
as a feedstock for synthesis gas in
methanol production 21
gasification 20
lignite and biomass, investment
costs of methanol production
from 35
Coating(s)
formulation, solvents in industrial
protective 79
systems, acrylic–melamine resin 81
systems, thermoset acrylic resin 81

Cobalt catalyst
carbonyl 73
ligand 73
system, oxo 160
Coconut oil 87
Cold crank simulator test 170
Conoco ALFOL alcohol process 93–95
Crank simulator test, cold 170
Crankcase lubricants, partial
synthetic automotive 170
diester vs. petroleum basestocks of 179
Crankcase oils 165
Crop residues 63
Cyclohexanol 182

D

DBP (dibutyl phthalate) 183
DCP (dicapryl phthalate) 182
DDG (distillers dried grains) 49
DDGS (distillers dried grains with
solubles) 49
Dehydration of alcohols 10–12
Denitrification, methanol as carbon
source for water 26
Denitrification of waste water 26
Detergency in dishwashing,
SAE 133, 138, 139f
Detergency, heavy-duty laundry
application, SAE 133, 134f,
135t, 136f
Detergent(s) 101–112, 113
ethoxylation of alcohols to 159
materials 163
olefin production 159
range alcohols 163
SAE performance in household ..131–140
Di-2-ethylhexyl phthalate
(DOP) 83, 181, 188
vs. DIDP in plastisol compounds,
economic comparison of ..190, 192t
Dibutyl phthalate (DBP) 183
Dicapryl phthalate (DCP) 182
DIDP (see Diisodecyl phthalate)
Diesel substitution, methanol as 44
Diester(s) 165
vs. petroleum basestocks of
partial synthetic automotive
crankcase lubricants 179
raw materials for 167t
for use in a partial synthetic
lubricant 166
Diethylaluminum hydride 93
Diethylene glycol mono-n-butyl ether
(butyl CARBITOL) 81
Diisobutylamine 83
Diisodecyl phthalate (DIDP) 187
in plastisol compounds, economic
comparison of DOP vs. 190, 192t

Diisoheptyl phthalate (Jayflex 77) 188
Diisooctyl adipate (DIOA) 170
Dimethyl terephthalate (DMT) 23
Dioctyl adipate (DOA) 182
Dioctyl azelate (DOZ) 182
Dioctyl sebacate (DOS) 182
Dioctylterephthalate (DOTP) 186
Diphenyl phosphate, octyl 182
Dishwashing liquids 112
alcohol ether sulfate, light-duty 107
Dishwashing, SAE
detergency in 133, 138, 139f
Distillers dried grains (DDG) 49
with solubles (DDGS) 49
DMT (dimethyl terephthalate) 23
DOA (dioctyl adipate) 182
DOP (see Di-2-ethylhexyl phthalate)
DOS (dioctyl sebacate) 182
DOTP (dioctylterephthalate) 186
DOZ (dioctyl azelate) 182

E

Energy
balance of methanol from wood
plant 34t
in Brazil, biomass 30
consumption in Brazil (1977),
primary 29t
crisis, impact of 20–22
Security Corporation programs for
biomass 57
Tax Act of 1978 57
Engine(s)
oil classification(s), passenger car.... 174
current API 176t
oil viscosity classification,
SAE low-temperature 175t
systems, compatibility ethanol–
gasoline blends with
automotive fuel 68
technical aspects of the use of
ethanol–gasoline blends in
spark ignition 64–69
text data on 5W–30 partial
synthetic motor oil 178t
Environmental effects of ethanol
production 63
EPAL alcohols 99
Epoxides, transformation of
chlorohydrins into 3
Epoxy stabilizer/plasticizers 184
Ester(s)
Bouveault–Blanc reduction of 207
epoxidized fatty acid 184
hydrogenation of methyl 89, 92f
manufacture 165–166
plasticizers for PVC, monohydric
alcohol 181–194

Ester(s) *(continued)*
  of SAE, sulfosuccinate half 149–152
  sulfosuccinate 113
Esterification 165–166
  alcohol 5–7
  of organic acids, acid-catalyzed 6
Ethanol 55
  as an automobile fuel 55
  biomass-derived 55
    1980–1995, potential demand
      for 59, 60*f*
    production, net energy gain in 63
  chemical and physical properties of 64
  cost of producing grain 59, 61
  fuel production, biomass feedstocks
      potentially available for 62*t*
  fuels policy issues 58–64
  –gasoline blends
    with automotive fuel/engine
      systems, compatibility of 68
    on engine emmission, effect of
      the use of 68
    in spark ignition engines,
      technical aspects of the
      use of 64–69
  in gasoline, octane-boosting
      properties of 64
  manufacture and applications 47–53
  in motor gasoline 55–69
  production
    availability of feedstocks for 59
    environmental effects of 63
    cellulose as a feedstock for 51
    by continuous fermentation of
      grain mash, commercial 53
    via hydrolysis of cellulose 50
    from sugar cane 31
    sulfuric acid ester process for
      synthetic 48
  as a suspect carcinogen 53
  use in the fragrance and flavor
      industry 202
Ether
  butyl glycol 81
  methyl tertiary butyl (MTBE) 65
  sulfate(s), alcohol 101, 107–111
    heavy-duty laundry powders 107
    light-duty dishwashing liquids 107
    shampoos 109
  sulfonates, alkyl glyceryl 102
  synthesis, Williamson 3
Ethoxylates, alcohol 101
  secondary (SAE) *(see* Alcohol
    ethoxylates, secondary)
Ethoxylates, nonylphenol (NPE) 117
Ethoxylation of alcohols to detergents 159
Ethoxysulfates 149
  from SAE 149
Ethyl alcohol 202, 204*f*

Ethyl process 96*f*
  -modified linear alcohol 95–99
Ethylene 88, 187
  alpha olefins, oligomerization of 159
  chain growth process, manufacture
    of higher straight-chain
    alcohols by 87–99
  Friedel–Crafts reaction of benzene
    and 207
  gas 47
  glycol mono-*n*-butyl ether (butyl
    CELLOSOLVE) 81
  oxide 109
    content in heavy-duty laundry
      powder, optimum 103, 104*f*
  production 159
  to triethylaluminum, addition of 91
2-Ethylhexanol (2EH) 181, 185
  applications of 83–84
  -based plasticizers 84, 182
  consumption pattern in the U.S. 80*t*
  diester of azelaic acid 174
  from propylene, production of 77, 79
  by the rhodium oxo process,
    manufacture 71–84
2-Ethylhexyl acrylate 84
2-Ethylhexyl glycolate, synthesis of 6

F

Fatty
  acid esters, epoxidized 184
  acids, hydrogenation of 89
  alcohols 87, 20*f*
    synthetic 88
  triglycerides 88
Feedstock(s)
  for alcohols, changing value of
    olefin 187
  for ethanol production, availability
    of 59
  methanol 21*t*
    naphtha as 20
    natural gas as 20
  potentially available for ethanol
    fuel production, biomass 62*t*
  for the production of ethanol,
    cellulose as 51
  for synthesis gas in methanol
    production, coal as 21
Fermentation
  alcohol demand in the U.S. 51
  alcohol manufacture 49–51
  continuous 51
    of grain mash, commercial
      production of ethanol by 53
  molasses 49
  and fragrance industry,
    monohydric alcohols in 197–210

Fermentation (*continued*)
  industry, ethanol use in the
     fragrance and 202
Floral extracts, number of alcohols
  found in natural 200*f*
Foam stabilizers 111
Formaldehyde 23
Fragrance(s) industry, monohydric
  alcohols in the flavor and 197–210
Friedel–Crafts reaction of benzene
  and ethylene 207
Fruit(s), alcohols in 199
  extracts 200*f*
Fuel(s)
  alcohol
    from biomass, government
     incentives (tax credits)
     to produce 56–58
    process economics, grain 59, 61
    from sources other than petro-
     leum, production of 56
  boiler 63
  from biomass, liquid 30
  engine systems, compatibility of
    ethanol–gasoline blends
    with automotive 68
  ethanol
    as an automobile 55
    policy issues 58–64
    production, biomass feedstocks
     potentially available for 62*t*
  methanol as 39, 43–44
    oil substitution 44
  motor 52
  uses of methanol 26

G

Gas
  as a methanol feedstock, natural ...20, 21
  to methanol, low vs. high
    pressure process to
    convert synthesis 22
  preparation, synthesis 19–20
  steam reforming of natural 21
Gasification
  charcoal 39
  coal 20
  wood 39
  methanol synthesis—conventional .. 33*f*
  methanol synthesis—Winkler 32*f*, 34
Gasohol 56, 65
  driveability problems with 66, 67*f*
  vs. gasoline 52
  impact of incentives cost of
    alcohol used in 62*t*
Gasoline
  blends, ethanol–
    with automotive fuel/engine
     systems, compatibility of 68

Gasoline (*continued*)
  blends, ethanol– (*continued*)
    on engine emissions, effect of
     the use of 68
    in spark ignition engines,
     technical aspects of the
     use of 64–69
  chemical and physical
    properties of 64
  ethanol in motor 55–69
  vs. gasohol 52
  high octane blending stock for 25
  octane–boosting properties of
    ethanol in 64
  octane improver 187
  substitution, methanol as 43
  unleaded 56
Geraniol 205, 206*f*
Grain(s)
  alcohol fuels process economics 59, 61
  distillers dried (DDG) 49
    with solubles 49
  ethanol, cost of producing 59, 61
  mash, commercial production of
    ethanol by continuous
    fermentation of 53

H

Halide(s)
  converting alcohols to alkyl 9
  methyl 23
  reaction of an alcohol with
    an acyl 6
  S_N2 displacement on an alkyl 3
Hemiacetals 7
Hexyl alcohol 184
Hydrocarbons, hydroxyl
  derivatives of 1–15
Hydroformylation of olefins to
  alcohols 159–163
Hydroformylation process 160
Hydrogen to methanol, conversion of
  CO and 15
Hydrogenation of fatty acids 89
Hydrogenation of methyl esters 89
Hydroxycitronellal 202, 204*f*
Hydroxyl
  derivatives of hydrocarbons 1–15
  group in alcohols, analytical
    determination of 14
  group in alcohols, displace-
    ment of 9–10

I

Isobutanol 73
  applications of 83
  consumption pattern in the U.S. 80*t*

Isobutyl acetate ... 83
Isobutyraldehyde ... 73
Isocyanates, reaction of
    alcohols with ... 8
Isodecyl alcohol ... 182
Isononyl alcohol ... 185
Isooctyl alcohol ... 182

J

Jayflex 77 (diisoheptyl phthalate) ... 188
Jones reagent ... 13

K

Kerosene, *n*-paraffins in ...115, 117*t*
Ketals ... 7
Ketones, reaction of alcohols with
    aldehydes and ... 7

L

Lacquers, nitrocellulose ...79, 82
LAS (linear alkyl-benzene
    sulfonate) ...109, 138
Laundry
    application, SAE detergency,
        heavy-duty ...133, 134*f*, 135*t*, 136*f*
    liquids, alcohol nonionic
        heavy-duty ...105–107
    powder(s)
        alcohol ether sulfate
            heavy-duty ... 107
        alcohol nonionic heavy-duty ...103–105
        optimum ethylene oxide content
            in heavy-duty ...103, 104*f*
Leaf alcohol ... 199
Ligand, catalyst cobalt ... 73
Lignite and biomass, investment
    costs of methanol production
    from coal ... 35
Lilac perfume ... 198*f*
Linalool ...205, 208*f*
Linear alkyl-benzene sulfonate
    (LAS) ... 138
Lubricant(s)
    base oils, miscibility of C$_{12\text{-}14}$
        SAE in various ... 144*t*
    basestocks from monohydric
        alcohols, synthetic ...165–179
    diesters for use in a partial
        synthetic ... 166
    partial synthetic automotive
        crankcase ... 170
    diester vs. petroleum
        basestocks of ... 179
    textile ... 141

M

Mash, clear ... 50
Mash, commercial production of
    ethanol by continuous
    fermentation of grain ... 53
Menthol ...205, 206*f*
1-Menthol ... 205
Melle process for the reuse of yeast ... 50
Methacrylamide with *n*-butanol,
    alcoholysis of ... 81
Methacrylate, *n*-butyl ... 81
Methanol
    to acetic acid, conversion of ... 25
    as carbon source for single-cell
        protein production ... 25
    as carbon source for water
        denitrification ... 26
    conversion of CO and H$_2$ to ... 15
    costs and oil prices ... 38
    current applications of ...22–25
    demand, world ... 24*t*
    feedstock(s) ... 21*t*
        naphtha as ... 20
        natural gas as ... 20
    as a fuel ...26, 39, 43–44
    low vs. high pressure process
        to convert synthesis gas to ... 22
    manufacture and uses ...19–26
    new applications of ...25–26
    plants, investment costs of ... 35*t*
    production ... 19
        coal as a feedstock for
            synthesis gas in ... 21
        from coal, lignite and biomass,
            investment costs of ... 35
        costs ... 37*t*
            total ... 38*t*
            plant, operating ...36–37
    as a substitution (for)
        diesel ... 44
        fuel oil ... 44
        gasoline ... 43
    synthesis ... 20
        —CESP's pilot plant ... 41*f*
        —conventional gasifier ... 33*f*
        process routes for ...31, 32*f*, 33*f*
        scheme ... 19*f*
        —Winkler gasifier ... 32*f*
    from wood
        in Brazil ...29–44
        experiments in the
            production of ...39–43
        plant, energy balance of ... 34*t*
Methyl
    ester hydrogenation ...89, 92*f*
    halides ... 23
    methacrylate ... 23
    tertiary butyl ether (MTBE) ...25, 65

Methylamines 23
Methylcyclopentadienyl manganese
    tricarbonyl (MMT) 64
Methylheptenone 205, 208*f*
Molasses fermentation 49
Motor oil(s)
    composition and laboratory test
        data on 5W-30 partial
        synthetic 177*t*
    engine test data on 5W-30
        partial synthetic 178*t*
    SAE viscosity classifications
        for 170, 172*t*
MTBE (methyl tertiary butyl ether) ..25, 65

N

Naphtha as a methanol feedstock 20
Naphtha, steam reforming of 20
Natural gas as a methanol feedstock .. 20
Natural gas, steam reforming of 21
Nitrocellulose lacquers 79, 82
Nonionics, alcohol 101
Nonylphenol ethoxylates (NPE) 117
Nucleophilic reaction of alcohols 2–9
Nucleophilic reactions of the
    alkoxide ion 3–4

O

Octane–boosting properties of
    ethanol in gasoline 64
Octane enhancers 64, 187
Octanol-2 (capryl alcohol) 182
Octyl diphenyl phosphate 182
Olefin(s) 161, 163
    to alcohols, conversion of 159–163
    to alcohols, hydroformyla-
        tion of 159–163
    alpha 159
        oligomerization of ethylene 159
    feedstocks for alcohols, changing
        value of 187
    process, shell higher (SHOP) 159–163
    production, detergent 159
α–Olefin(s) 91, 97
Olefinic alcohols 199, 20*f*
Oil(s) 88, 170
    additives, lube and fuel 84
    classification(s), passenger car
        engine 174
        current API 176*t*
    coconut 87
    crankcase 165
    entitlements program 57
    epoxidized soybean 184
    miscibility of C_{12-14} in various
        lubricant base 144*t*

Oil(s) (*continued*)
    motor
        composition and laboratory test
            data on 5W-30 partial
            synthetic 177*t*
        engine test data on 5W-30
            partial synthetic 178*t*
        SAE viscosity classifications
            for 170, 172*t*
    partial oxidation of residual 21
    petroleum-based 174
    prices, methanol costs and 38
    substitution, methanol as a fuel 44
    viscosity classification, SAE low-
        temperature engine 175*t*
Organic acids, acid-catalyzed
    esterification of 6
Organic acids, reaction of alcohols
    with 5–7
Oxalic acid, process to make 15
Oxidation of alcohols 12–14
Oxidation, partial 20
    of residual oil 21
Oxide, ethylene 109
Oxides, amine 102, 112
    alkyl dimethyl 102
Oxo
    alcohols 88
        C_4 79
        manufacture, higher linear ...159–163
    catalyst(s) 73
        system 160
        cobalt 160
    process, rhodium 182, 183
        -catalyzed 185
        flow diagram of 75, 76*f*
        manufacture of *n*-butanol 71–84
        manufacture of 2-ethylhexanol
            by 71–84
Oxonation process(es), propylene 71
    comparison of 72*t*
Oxonation, rhodium low-pressure 73–75

P

PAE (primary alcohol ethoxylates) 117
PAO (polyalphaolefin) 166
Paper industry, SAE uses in the
    pulp and 145
Paper-making, sulfite process for 50
Paraffins 115, 161
n-Paraffins 88
    in kerosene 115, 117*t*
Perfume(s) 199
    lilac 198*f*
Petroleum 166
    -bases oils 174

Petroleum (*continued*)
basestocks of partial synthetic
automotive crankcase
lubricants, diester vs. 179
Phenolic resins 23
Phenylethyl alcohol 199, 207, 209*f*
Phosphoric acid 48
Phthalates 184
in automotive upholstery, linear 187
monohydric alcohol 181–194
Pinner synthesis 8
Plasticizer(s)
cellulosic 181
cost–performance of 188–190
epoxy stabilizer/ 184
2-ethylhexanol-based 84, 182
fast-solvating 188
lower volatility 190
polyvinyl chloride (PVC)
intermediates for 184
resins 82
-range alcohols 163
Plastisol compounds, economic
comparison of DOP vs.
DIDP in 190, 192*t*
Polyalphaolefin (PAO) 166
Polymerics 182
Polyvinyl chloride (PVC)
plasticizers 83
intermediates for 184
monohydric alcohol ester 181–194
resins 82
Powders, alcohol ether sulfate
heavy-duty laundry 107
Propylene 187
oxide adducts of SAE 146
oxonation process(es) 71
comparison of 72*t*
production of 2-ethylhexanol
from 77, 79
Protein production, methanol as
carbon source for single-cell 25
Pulp and paper industry, SAE
uses in 145
Pulp using nonionic surfactant,
deresination of 147*t*

R

Resin(s)
amino 23
systems 81
coating systems, thermoset acrylic .. 81
coating systems, acrylic–
melamine 81
phenolic 23
plasticizers for PVC 82
Rhodium low-pressure oxonation 73–75

Rhodium oxo process
-catalyzed 185
flow diagram of 75, 76*f*
manufacture of *n*-butanol by 71–84
manufacture of 2-ethylhexanol by ..71–84
River die-away test for biodegradability of SAE 124–127
Ritter reaction 10
Rose alcohols 205
Rose base 198*f*

S

S$_N$2 displacement on an alkyl halide .. 3
Saccharomyces 51
SAE (*see* Alcohol ethoxylates,
secondary)
Sandalore 207, 209*f*
Secondary alcohol ethoxylates (SAE)
(*see* Alcohol ethoxylates
secondary)
Sesquiterpenic alcohols 202, 203*f*
Shampoos, alcohol sulfate 111
ether 109
SHOP (shell higher olefin process) .159–163
Sludge test for biodegradability of
SAE, continuous-flow
activated 127, 128*f*
Sodium citrate 105
SOFTANOL 113
Solvents 25
in industrial protective coating
formulations 79
Soybean oil, epoxidized 184
Spark ignition engines, technical
aspects of the use of ethanol–
gasoline blends in 64–69
Stabilizer/plasticizers, epoxy 184
Steam reforming 20
of naphtha 20
of natural gas 21
Stillage, thin 63
Sugar(s)
cane, ethanol production from 31
for conversion to alcohol,
conversion of cellulose
to fermentable 52
degradation of cellulose to 50
Sulfate(s), alcohol 111
ether 101, 107–111
light-duty dishwashing liquids 107
heavy-duty laundry powders 107
shampoos 109
shampoos 111
Sulfates, alkyl 102
Sulfonates, alkyl glyceryl ether 102, 112
Sulfonates, alkylbenzene (LAS) 107, 109

Sulfite process for paper-making 50
Sulfosuccinate esters 113
Sulfosuccinate half esters of SAE ..149–152
Sulfuric acid ester process for
    synthetic ethanol production 48
Surfactant(s) 113
    applications of higher alcohols
        in household 101–112
    deresination of pulp using
        nonionic 147*t*
Synthesis gas
    to methanol, low vs. high pressure
        to convert 22
    in methanol production, coal as a
        feedstock for 21
    preparation 19–20
Synthetic alcohol
    demand in the U.S. 51
    fatty 88
    manufacture 47–49
Synthetic ethanol production, sulfuric
    acid ester process for 48

T

Tallow 87
Tax
    Act of 1978, Energy 57
    bill, Windfall Profits 56
    credits to produce alcohol fuels
        from biomass, government
        incentives 56–58
TBA (tertiary butyl alcohol) 65
TCP (tricresyl phosphate) 182
Terg-o-Tometer tests 133
TERGITOL 15-S 113
Terpenoid alcohols 202, 203*f*
Tertiary butyl alcohol (TBA) 65
Tetrapotassium pyrophosphate
    (TKPP) 105
Textile industry, SAE uses in 141–145
Textile lubricants 141
Thin stillage 63
Thymol 205
TKPP (tetrapotassium
    pyrophosphate) 105
TMA (trimellitic anhydride) 184
TOF (trioctyl phosphate) 182
Tosylates 10
Toxicity to fish, SAE 127–131
Trialkylaluminum compounds 94, 97
    detergent-range 98

Trichoderma reesei 50
Tricresyl phosphate (TCP) 182
Triethylaluminum, addition of
    ethylene to 91
Triethylaluminum, synthesis of 90, 93
Triglycerides, fatty 88
Trimellitic anhydride (TMA) 184
Trioctyl phosphate (TOF) 182
Triphenylphosphine 74

U

Upholstery, linear phthalates
    in automotive 187
Upholstery, vinyl 187

V

Vinyl upholstery 187
Viscosity classifications for motor
    oils, SAE 170, 172*t*

W

Water, denitrification of waste 26
Williamson ether synthesis 3
Windfall Profits Tax bill 56
Winkler-type gasifier 34
    methanol synthesis— 32*f*
Wood
    costs and productivity 36
    gasifications 39–43
    methanol from
        in Brazil 29–44
    experiments in the produc-
        tion of 39–43
    plant, energy balance of 34*t*

Y

Yeast, Melle process for the
    reuse of 50

Z

ZDDP (zinc dialkyl
    dithiophosphates) 83, 84
Ziegler alcohols 88
    production of 90–93
Zinc dialkyl dithiophosphates
    (ZDDP) 83, 84
Zymomonas 51